The world railway system

The world railway system

BERNARD DE FONTGALLAND FCIT
Honorary Secretary General of the International Union of Railways

The right of the University of Cambridge to print and sell all manner of books was granted by Henry VIII in 1534. The University has printed and published continuously since 1584.

CAMBRIDGE UNIVERSITY PRESS
Cambridge
London New York New Rochelle
Melbourne Sydney

Published by the Press Syndicate of the University of Cambridge
The Pitt Building, Trumpington Street, Cambridge CB2 1RP
32 East 57th Street, New York, NY 10022, USA
296 Beaconsfield Parade, Middle Park, Melbourne 3206, Australia
Originally published in French as *Le système ferroviaire dans le monde* by Editions Celse

First published in English by Cambridge University Press 1984 as *The world railway system*

Printed in Great Britain at the University Press, Cambridge
Library of Congress catalogue card number: 84-1820

British Library cataloguing in publication data
Fontgalland, Bernard de
The world railway system.
1. Railroads–Management
I. Title II. Le système ferroviare dans
le monde. *English*
385'.068 TF505

ISBN 0 521 24541 9

PN

CONTENTS

FOREWORD TO THE FRENCH EDITION

It is a tradition of Secretaries General of the International Union of Railways to enhance the intellectual literature of our profession; with this book Bernard de Fontgalland confirms to a wider audience his place within that highly-respected tradition. His work is particularly timely in its appearance in a world where, especially for us in the economically-developed nations, the dangers of the profligate use of limited resources become more acute as each day passes. For railwaymen these dangers present both a challenge and an opportunity. We are in the debt of Monsieur de Fontgalland for the rigour and precision of his analysis.

The challenge is to adapt our industry to the needs of a world that can no longer afford to waste its non-renewable resources, the opportunity is that our industry should take its rightful place in what I have called the commonwealth of transport. Monsieur de Fontgalland rightly emphasises – and I urge all my fellow railwaymen to keep this point ever with them – that transport is not an end in itself, but only a means to an end. In particular he underlines the role played by the state and its agencies – even in the most liberal of economies – in shaping and controlling those ends, and hence those of our industry.

If I may inject a national note in discussing a work of such international perspective, I believe my British colleagues will draw some satisfaction from the author's endorsement of the correctness of their approach and the wider relevance of their work. They have faced a decade of the fiercest doubt about the role of the railways. Their sympathy will be extended to those of their colleagues in other countries who are now similarly confronted by the imperatives of change, and who have still some way to go down the track. But, as Monsieur de Fontgalland points out, pruning a tree is realistic because it ensures its health and future growth. It is his vision of the future that forms for me the abiding message of this book.

It is for this generation of railwaymen to so inspire that vision in our respective communities that we generate the will and the means to bring it to reality. This book puts us all, in the world-wide railway community, permanently and even more deeply in debt to Bernard de Fontgalland, railwayman extraordinaire.

Peter Parker
15th January 1979

ABBREVIATIONS

AAR	Association of American Railroads (Washington)
ALAF	Asociación Latino-Americana de Ferrocarriles Latin American Association of Railways (Buenos Aires)
Amtrak	American Transportation on Track or National Railroad Passenger Corporation (Washington)
BR	British Rail (London)
CCI	Chambre de Commerce Internationale (Paris) International Chamber of Commerce
CCITT	Comité Consultatif International Télégraphique et Téléphonique (Geneva) International Consultative Committee on Telecommunications
CEA	Commission Economique (des Nations Unies) pour l'Afrique (Addis Ababa) United Nations Economic Commission for Africa
CEE	Commission Economique (des Nations Unies) pour l'Europe (Geneva) United Nations Economic Commission for Europe
CFF	Chemins de Fer Fédéraux suisses (Berne) Swiss Federal Railways
CGST	Conception Globale Suisse des Transports
CIM	Convention Internationale concernant le transport des Marchandises par chemin de fer International Convention on Goods Transport by Rail
CIT	Comité International des Transports par chemin de fer (Berne) International Committee on Railway transport
CIV	Convention Internationale concernant le transport des Voyageurs et des bagages par chemin de fer International Convention on Transport of Passengers and Luggage by Rail
CIWL	Compagnie Internationale des Wagons-Lits (Paris)
CMO	Chemins de fer du Moyen-Orient Tariff union for the railway networks of Turkey, Syria, Lebanon, Iraq and Iran
CMEA	Council of Mutual Economic Assistance (Moscow)

CN Canadian National Railway (Montreal)
COFC Container on flat car
Conrail Consolidated Rail Corporation (Philadelphia)
CP Canadian Pacific Railway (Montreal)
DB Deutsche Bundesbahn
German Federal Railways (Frankfurt)
DOT Department of Transportation (Washington)
ECMT European Conference of Ministers of Transport (Paris)
EEC European Economic Commission
Eurailpass European railway season ticket for non-Europeans
Eurofima Société Européenne pour le Financement du Matériel ferroviaire (Basle)
European Society for the Financing of Railway Rolling Stock
FIATA Fédération Internationale des Associations de Transitaires et Assimilés (Zurich)
International Federation of Associations of Forwarding Agents and Equivalents
FRA Federal Railroad Administration (Washington)
FUAAV Fédération Universelle des Associations d'Agences de Voyages (Brussels)
Universal Federation of the Association of Travel Agents
GCTM Gestion Centralisée du Trafic Marchandises
Centralised Management of Goods Traffic
Gosplan Gosudarstvennyi Planovyj Komitet
State Planning Committee (Moscow)
ICC Interstate Commerce Commission (Washington)
IMF International Monetary Fund (Washington)
Interrail Special ticket for people of 25 or under valid on 20 European networks of the UIC
IRCA International Railway Congress Association (Brussels)
IRU International Road Transport Union (Geneva)
ISO International Standards Organisation (Geneva)
JNR Japanese National Railways (Tokyo)
KCR Kowloon–Canton Railway (Hong Kong)
KNR Korean National Railroad (Seoul)
MIS Management Information System
MOS Management Optimisation System
MTT Mezdunarodnyj transitnyj tarif
International transit tariff
N de M Ferrocarriles Nacionales de Mexico
National Railways of Mexico
NS N.V. Nederlandse Spoorwegen
Netherlands Railways (Utrecht)
OCTI Office Central des Transports Internationaux par chemin de fer (Berne)
Central Office of International Railway Transport
OPEC Organisation of Petrol Exporting Countries
OSZD Organizacija Sotrudnicestva Zeleznyh Dorog (Warsaw)
Organisation for Railway Collaboration
PARCA Pan American Railways Congress Association (Buenos Aires)
Ra-Ra Rail-on rail-off (transferring wagons from rail to ship)
RIC Regolamento Internazionale Carrozze
International Union of Coaches and Luggage Vans (UIC)

RIV	Regolamento Internazionale Veicoli International Union of Wagons (UIC)
RMS	Railway macrosystem
Ro-Ro	Roll-on roll-off (transferring road vehicles from road to ship)
SAR	South African Railways
SMGS	Soglasěnie o Mezdunarodnom Gruzovom železnodorožnom Soobsčěnii Convention on International Railway Goods Traffic
SMPS	Soglasěnie o Meždunarodnom Gruzovom zěleznodorožnom Soobsčěnii Convention on International Railways Passenger Traffic
SNCF	Société Nationale des Chemins de fer français (Paris) French National Railways
STOL	Short take-off and landing
TANZARA	Tanzania Zambia Railway Authority (Dar-es-Salaam)
TEE	Trans-Europ-Express
TEEM	Réseau de trains Trans-Europ-Express-Marchandises Network of Trans-Europ-Express goods trains
TGV	Train à grande vitesse French high-speed train
TMS	Tricontinental megasystem
TOFC	Trailer on flat car
TOPS	Total Operation Processing System
TRACS	Traffic Reporting and Control System
TRAIN	Tele-Rail Automated Information Networks
Transfesa	Société de transports Ferroviaires speciaux (Madrid)
UAR	Union of African Railways (Kinshasa)
UIP	Union Internationale des associations de propriétaires de wagons de Particuliers (Geneva) International Union of Associations of Private Wagon Owners
UIRR	Union Internationale des sociétés de transport combiné Rail/Route (Frankfurt am Main) International union of Combined Rail-Road Transport Companies
UITP	Union Internationale des Transports Publics (Brussels) International Union of Public Transport
UNO	United Nations Organisation (New York)
USRA	United States Railway Association (Washington)
VIA	VIA Rail Canada Inc. (subsidiary of CN which runs all rail passenger services in Canada except suburban trains) (Montreal)
VTOL	Vertical take-off and landing

INTRODUCTION

The railway is a mode of transport, or more precisely, a land-based guided form of transport, which is dependent on the tertiary industries of the economy. Transportation is in fact a service for the benefit of the whole community, not an end in itself. In the course of history the range of methods of transport has widened considerably yet it is remarkable that none of the early methods has been completely eliminated by the more modern ones. Though enormous technological advances have been made, and the markets have become more specialised the powerful incentive to invention of competition has made sure that this essential part of the economy remains open and dynamic.

The railway is a very recent form of transport. The dawn of transport by water may be traced to the time of Noah's Ark, although then of course it had no infrastructure. The network of Roman roads was the first land-based system of transport to combine vehicles with a specialised infrastructure. The railway, however, has only just celebrated the 150th anniversary of the first passenger train between Stockton and Darlington in 1825, having originally been invented for the transport of coal only a few decades earlier.

The pipeline was developed in the US from 1865 onwards. The second age of road transport, which began with the invention of the internal combustion engine and the pneumatic tyre, dates from the end of the nineteenth century, and that of water transport, marked by the replacement of the sail by the steam engine, dates from the same time. Air transport really began around about 1920 and has just reached its 50th birthday. Since then no new system of transport, particular no land-based guided form, has come into operation except as an experiment or a one-off design.

For several thousand years there were no radical changes in the various forms of transport, then in just one century we saw the birth and development of all the present systems. This is the century of the industrial revolution, with which the railway remains intimately associated. For it is the railway which made this revolution possible, in Europe and the US from the beginning, then later in Russia and Latin America, and finally in Asia and Africa. The railway boom now occurring in the Middle East is the most recent and eloquent affirmation of this association.

However short, the history of the railway has had its ups and downs. Its first century was marked by continuous capitalistic expansion. Governments, whatever their economic policy at the time, soon felt it necessary to check the monopoly won by the railways in only a few decades by imposing various concessions or 'obligations of public service' on the private railway companies. These were intended to assure a measure of protection to its users as well as to safeguard national, political, economic and military needs.

From the end of the nineteenth century, and so before the advent of the modern car, some difficulties did, however, emerge in Europe. The first analyses of the structure of costs on the railway highlighted the preponderance of fixed charges, in particular expenditure on staff. This was at a time when automation was as yet unknown and electric or diesel traction were still in the design stages. These analyses showed the intrinsic non-profitability of numerous lines of recent construction which had little traffic. And so it became necessary for the state to set up a system of subsidies, or else buy back lines and networks which were in deficit, which eventually came to the same thing. But no clear economic policy was devised then, let alone implemented. Events after the First World War multiplied these difficulties, which were further increased by the economic crisis of the thirties.

At the same time, automobile transport, individual and collective, developed very quickly in almost total freedom and creamed off a lot of railway traffic. This tolled the death knell for the railway's century of monopoly. The worsening financial position of railway companies led some governments to set up policies of 'rail–road coordination' with respect to regulations, financing and price control. But these efforts did nothing to alleviate the basic problems, such as lack of awareness of costs to the community or the choice of investments. Other countries preferred to nationalise their railways. So in highly industrialised countries the expansion of the railways was to all intents and purposes halted from 1930 onwards, as was the modernisation of existing infrastructures.

In developing countries, however, the construction and operation of railways by the state was the rule, the railway being thought of as an essential tool of the economy, though unprofitable in the short term.

If the Second World War was to confirm the indispensable role of the railway, it nonetheless left the networks in a disastrous state of destruction or at the very least suffering from abnormal wear and tear through lack of renewal. With the exception of the US, the last big private companies disappeared in the upheaval. After the end of the war, particularly in Europe, most governments made the restoration of the railways one of their priorities. The need to work quickly meant that the reconstructed networks were identical to the previous ones: lines were rebuilt with the same characteristics as before and the first enormous orders for rolling stock were almost totally of the same type as those used before the war. So between 1950 and 1955, when reconstruction was virtually completed, the railway had the same infrastructure as during the nineteenth century.

It was very different with the other forms of transport. The accelerated development of the car industry led to a gradual doubling of the main road axes through the building of motorways, which constituted an enormous revolutionary change in capacity, speed and safety. Commercial aviation, benefiting from enormous technological strides brought about by the war, developed very quickly. In just a few years airlines robbed the railway of a considerable part of its market for medium-distance passengers in Europe, and nearly the whole market in North America. Air travel had a modern infrastructure which was constantly improved, culminating in the giant airports of today, some of which cover the same area as a large town. Pipelines were at first limited to regions which produced crude oil – the US, USSR and the Middle East – but in their turn they invaded the consumer countries, while increase in demand led to the construction of pipelines for refined oil products, which up to then had been an important market for the railways. Finally, some countries built or modernised their waterways so that they could carry a large amount of goods traffic, such as the Moselle and the St. Lawrence, and these cut deeply into the heavy goods market that had originally given rise to the railway. With the exception of pipelines these infrastructures were financed either wholly or in part by public funds, and their use remained exempt from the obligations of public service which continued to burden the railways.

Thus the railway, in countries with liberal economies, entered a black period in the decade from 1955 to 1965. Its importance as a transport system steadily decreased, the deficits of the railways increased, and many economists predicted the unavoidable decline if not the death of the railway, except for the strictly captive sectors of public service, such as urban and suburban passenger transport. The railways were unable to make investments in capacity and, more importantly, to renew the infrastructure of their major routes.[1] All that remained possible were strictly controlled investments in productivity, with a view to reducing the deficit by electrification, dieselisation and automation.

Countries with planned economies, however, never doubted the essential role of the railway in the community, and constantly implemented policies of investment in capacity and productivity. The most striking example is shown by the Soviet system which with 10 % of the route length carries more than 55 % of the world's goods traffic, with passenger traffic also extremely important. The developing countries too suffered no period of stagnation in the construction of their railways either.

However it was during this black period that a new kind of railway was conceived, primarily in Western Europe and Japan, a railway which would quickly reverse the tendency to think of the railway as nothing but a passing moment in the long history of transport. There are many reasons for this renaissance, but they are all based directly or indirectly on an economic analysis emanating from influences outside the railways – a circumstance which ensures the objectivity of the analysis. Here is a brief list of the main reasons for this renaissance; they will be examined in more detail in the course of this book:

(*a*) the determination and comparison of the aggregate costs to the community of the different modes of transport, i.e. investment and running costs;
(*b*) the special adaptability of the railways to cybernetics;
(*c*) the comparative technological evolution of the different forms of transport;
(*d*) their differing ecological impact;
(*e*) the development of the idea of the transport chain and of intermodal techniques.

The railways once more benefited from substantial investments, their traffic increased (beating all records in 1974), and, without doubt most significantly of all because it looked to the future, construction of new intercity lines started again, first in Japan and later in Europe. At the same time, motive power and operating methods were transformed and automation was brought into all departments of the railway, considerably reducing costs.

If we take a look back into the past, it is hardly surprising that the railway changed more in the two decades from 1955 to 1975 than it did in the century from 1840 to 1940. It owes this extremely rapid change to a generation of great railwaymen who were passionately devoted to the objective of providing the best possible service to the community, and so able to inspire multi-disciplinary teams to be more outwardlooking. Among those in the first rank we must pay homage especially to Louis Armand, Boris Pavlovich Beshehev, Franciscus den Hollander, Heinz-Maria Oeftering, Alfred Perlman and Hideo Shima.

It is only now that the railway has finally attained its full international dimension. Political and economic changes brought about by the last war and increasing worldwide cooperation in many fields of human activity have led to a multinational approach to the problems affecting land-based transport of the kind air and sea transport have always had. Intergovernmental bodies have been created and organisations which integrate the rail networks have taken on increasing importance. With the exception of very specialised lines (notably urban or mining lines), no railway system can take an important decision in complete isolation, even if it is not at present linked to other systems, because it might compromise its future.

It is this new railway, whose renaissance is already taking it into the twenty-first century, which I intend to describe here in its essential aspects. I shall do this using the concept of system as my guideline. By 'system' I mean 'the sum of the elements in dynamic interaction, organised as a function of a goal'.[1] This book is, in summary, about the application of the methods traditionally used in 'dynamic systems analysis' to the railways. Because the railway is a guided form of transport, such an approach is particularly appropriate. For in cybernetics we have both a methodological tool which is common to every enterprise, and an operational tool which is very specific to the railway.

1

Principles, parameters, systems

11 Principles

The railway is the result of the association of two physical principles: the guidance of wheel on rail and the adhesion of steel on steel; and of an operational technique, the convoy.

111 *Guiding the wheel on the rail*

This is a very old idea, first used in the sixteenth century mining works. Guiding is effected by wheels which have been fitted with a flange no more than a few centimetres in size. This extremely simple system has proved suitable to allow, in complete safety, the continuous increase in speed demanded by the development of the market. The 300 km/h threshold, which was reached in 1955, does not in any way constitute a technological 'barrier'. Speeds in excess of 1000 km/h have actually been obtained with experimental engines, but only over very short distances and for purely scientific purposes.

Railway guiding is on two rails. The monorail may seem simpler, and ever since the beginning of the railway it has excited the imagination of inventors. However it has always met with one fundamental difficulty, and that is a geometrical problem: how to design the points. By definition, a guided system can only be operational if the convoys can change track, whether they be working normally or whether it becomes necessary as a result of some traffic incident. The points must provide the same level of safety as the open line, and they must be easily manoeuvrable so that they can be crossed at as near as possible to the speed of the main line when they are switched. Also their cost and maintenance must be as low as possible. In practice these different conditions are not compatible with the geometry of the monorail and vehicle couple, and they explain the failure of all attempts to introduce monorail systems, including the recent so-called non-conventional

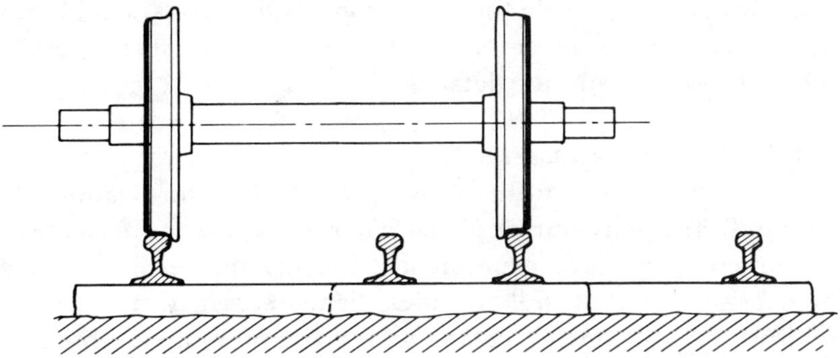

Fig. 1. Guiding and points

forms of guided transport which do not depend on rail–wheel contact but use the lift given by fluid or magnetic flux. The few monorails in current use in the world are isolated lines just a few kilometres in length, carrying urban or suburban passenger traffic and running infrequently[1]. To be more precise, the railway can be defined as 'birail with rail–wheel guidance by flanges under the body of the vehicle'. The space above the level of the rails is clear up to the level of the axle, that is about 30 cm. This leaves room for shallow bars to be placed between the two rails to collect the electric haulage current, in special cases, for cables or rack-rails; and then there are various information and safety devices (figure 1).

While guiding implies a rigid system, it is the points system which gives the necessary flexibility to railway operation by making it possible to create 'reservoirs' of equilibrium, made up by the group of tracks in sidings. A simple siding consists of another track run in parallel to a single track. The most important sidings, which are found in marshalling yards, may contain up to 50 tracks. Optimising the layout and use of sidings is one of the essential objectives of railway operation.

A major result of guiding is that the vehicles' course will be linear. A knowledge of the geography of the track is enough to define a complete route. Methods of transport which have physical guidance, such as railways and pipelines, function in one-dimensional space, while roads and waterways need two and aviation three. Otherwise the choices are binary: the points system is either to the right or to the left; a section of track is either free or occupied. We can therefore apply Boolean algebra, which is one of the bases of cybernetics.

In the operational field this factor constitutes a considerable advantage as compared to other forms of transport, for it makes it possible to schedule traffic in minute detail.

112 *Steel-on-steel adhesion*

'Adhesion is to the railway what lift is to aviation' (Louis Armand). This is a complex physical notion, which has been the object of numerous research experiments, notably in the development of high speeds, but it still includes elements which are difficult to interpret. Adhesion is a component of rolling motion which makes possible the transmission of traction and braking efforts. It results from two factors in steel-rail-on-steel-wheel contact:

(*a*) the resistance to forward motion, which manifests itself in motion by a very small amount of friction – about 1 kg per tonne. By way of comparison, on a road the contact between hard surface and tyre leads to friction ten times greater[1].

(*b*) traction, which involves sliding friction due to the relative displacement of the surfaces in contact.

On stopping, rail–wheel contact is effected according to a 'Hertz ellipse' of variable dimensions, and it is possible to calculate the stress produced in the different parts of the ellipse. But as soon as there is any movement, very different physical phenomena are observed on the surfaces in contact. Different theories have been put forward, but establishing their validity comes up against the near-impossibility of undertaking and setting up experiments. Even so, we know that a wheel cannot transmit a force, acceleratory or retarding, without sliding, and that it is possible to develop a maximal tangential stress before a wheel begins turning on the same spot (slipping). The relationship between this maximal tangential stress F and the pressure of the wheel on the rail Q, called the traction or braking coefficient, is variable. Its maximum value is called the coefficient of adhesion $\mu = F_{max}/Q$. Figure 2 shows the development of the phenomenon. An initial region of pseudo-sliding can be distinguished, then a region of micro-sliding up to the maximum value of μ. This occurs at 1–3 % of sliding on a dry rail, but may reach 5–8 % on a wet or greasy rail. Further, when the traction coefficient is lowered, it is the macro-sliding region which leads to slipping. The same curve is found for the movement of a tyre on a hard surface; the maximum is then reached at 15–20 % of sliding.

The coefficient of adhesion varies as a function of several parameters – state of the surface, weight of the rail, gradient of the track, and above

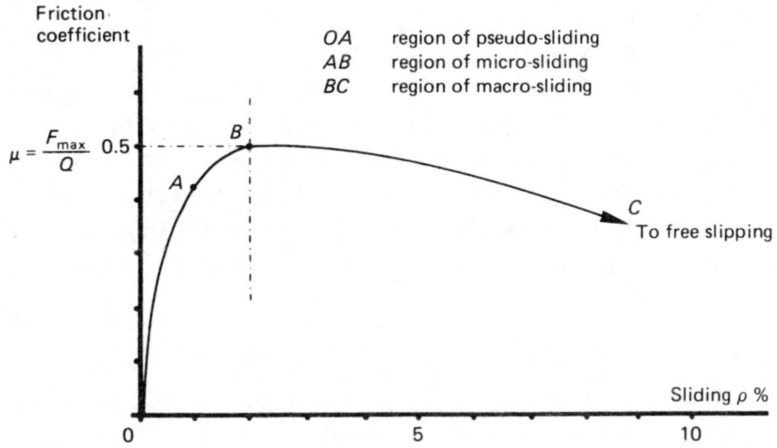

Fig. 2. The adhesion coefficient

all, speed – over a range of values between 0.1 and about 0.5. Networks adopt values for μ depending on their own particular criteria. These generally distinguish a starting coefficient and a running coefficient. Various devices are used to increase the value of μ (sanding, cleaning the rail, etc.). However recent research has shown that traction motors, and notably their regulation systems, affect adhesion and out of this have come some constructive new developments. Finally, it has been proved that adhesion remains sufficiently great at very high speeds (300 km/h) to make it possible to develop the power required to reach and maintain these speeds.

113 *The convoy*

A convoy is made up of powered and hauled vehicles which are coupled together but can also be uncoupled. This technique has several advantages. The composition of the convoy can be modified according to demand and during a journey; engines and hauled vehicles can be standardised independently; and staff numbers are very low in relation to the load carried. On the other hand, convoys require complex manoeuvres in making up a train and in shunting.

Operation of convoys is especially simple for a guided system. The railway convoy is the train, and its length is extremely flexible. We shall see later that improving the structure of trains is one of the essential objectives of railway production. The advantages of the convoy are so great that non-guided transport systems endeavour to make use of it – river navigation via the pushing technique, roads via tractors and trailers – but these are limited developments.

12 Parameters

The fundamental parameters of the railway system are numerous and rarely independent of one other. In addition the dividing line between parameters and given particulars is not fixed but varies according to whether we are talking about constructing a new system or operating an existing one isolated or linked to other systems. It is useful to list them as completely as possible here at the beginning of this book and develop each as necessary in the different chapters that follow.

121 *Geometrical parameters*

Essentially there are four of these: maximum gradient; minimum curvature on bends; the gauge of the track; and the loading gauge.

The first two are not independent. They are closely linked to topography, to an operational parameter, speed; and to an economic factor, the cost of construction. Fixing these parameters can only be the result of compromise, worked out for sections of line that are as long as possible. The railway is thus subject to particularly severe constraints. A gradient greater than 1.5 % considerably increases the installed power required, and only on major routes crossing mountain ranges such as the Alps or the Andes will railways accept gradients of 2.5–3 %. Large numbers of viaducts and tunnels and reducing the radius of curves considerably increases the cost of haulage. Topographical relief therefore imposes a fundamental limit on the diffusion of the railway as compared to the road for which gradients nearly ten times greater and curves of only a few metres radius are possible, which means that the road can go almost anywhere. However, the railway has an advantage over waterways, which require virtually horizontal stretches of water. Problems of topography can be circumvented by various measures, such as the use of cables or racks on tourist lines high up in the mountains or to cross cliffs impossible to go around, e.g. those between Sao Paolo and Santos. These are exceptions however, and they cost a great deal.

The gauge of the track became a given factor after a certain number of values had been standardised; however it may constitute a parameter in some projects for new lines. There are two familiar types of gauge: 'standard gauge' (1.435 m), and broad gauge' (1.524 m, 1.600 m, 1.676 m). These represent about 80 % of the total length of the world network. Any difference is the result of technical or political choice. The metric gauge (1 m and 1.067 m), which is used by the remaining

20 %, is essentially to be found in the developing areas: South East Asia; Latin America and the Southern Sahara. Their adoption is chiefly the result of economic considerations, for the metric gauges give the same operational possibilities of standard and broad gauges, with the sole exception of high speed.

Simple devices for changing gauge have been developed, so gauge conversions are no longer of economic value except in exceptional cases. In effect, the sole determining criterion for a new line is its compatibility, present or future, with adjacent networks, so as to avoid heavy investments and operational problems not compensated by real advantages.

The loading gauge (Figure 3) is the cross section inside which the vehicles and their loads must remain in order to preclude any possibility of collision with fixed obstacles or passing trains. We may distinguish:

(*a*) the loading gauges for rolling stock, both the static, which is defined by the cross section of a stationary vehicle, and the kinetic, which takes into account the actual displacements of vehicles and thus includes a safety margin.

(*b*) the clearance gauge, which is calculated from the former factors by adding a safety margin to facilitate the conveyance of exceptionally large loads.

Increasing speeds have led to this margin being increased to allow for shock waves produced by two convoys passing in opposite directions.

The loading gauge is derived from the dimensions of horse-drawn vehicles superseded by the rise of the railway. It must allow room for a human being to stand, move around and lie down crosswise in the vehicle, while at the same time leaving room for passage. Taking into account the dimensions of the wheels, this gives a rectangle approximately 3 m wide and 4.50 m high, realisable whatever the gauge of the track. The development of intermodal techniques has shown up the importance of the dimensions of the loading gauge.

122 *Topological parameters*

Railway systems are dependent on several topological structures (figure 4). Linear systems consist of isolated lines. There is only one route to follow. Typical cases are mining lines or metropolitan subways. The linear structure was the first stage in the construction of railway systems.

Branch systems were developed from the linear structure, either in

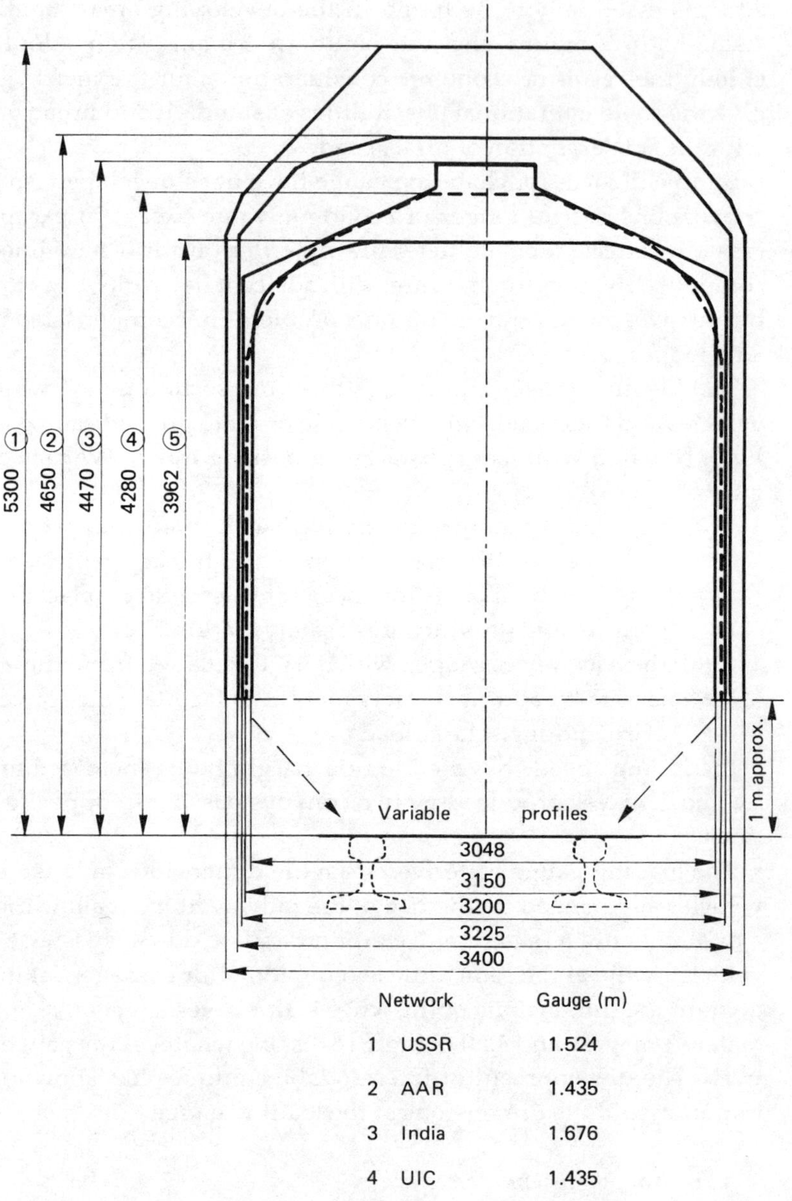

Network		Gauge (m)
1	USSR	1.524
2	AAR	1.435
3	India	1.676
4	UIC	1.435
5	SAR	1.067

Fig. 3. Principal loading gauges

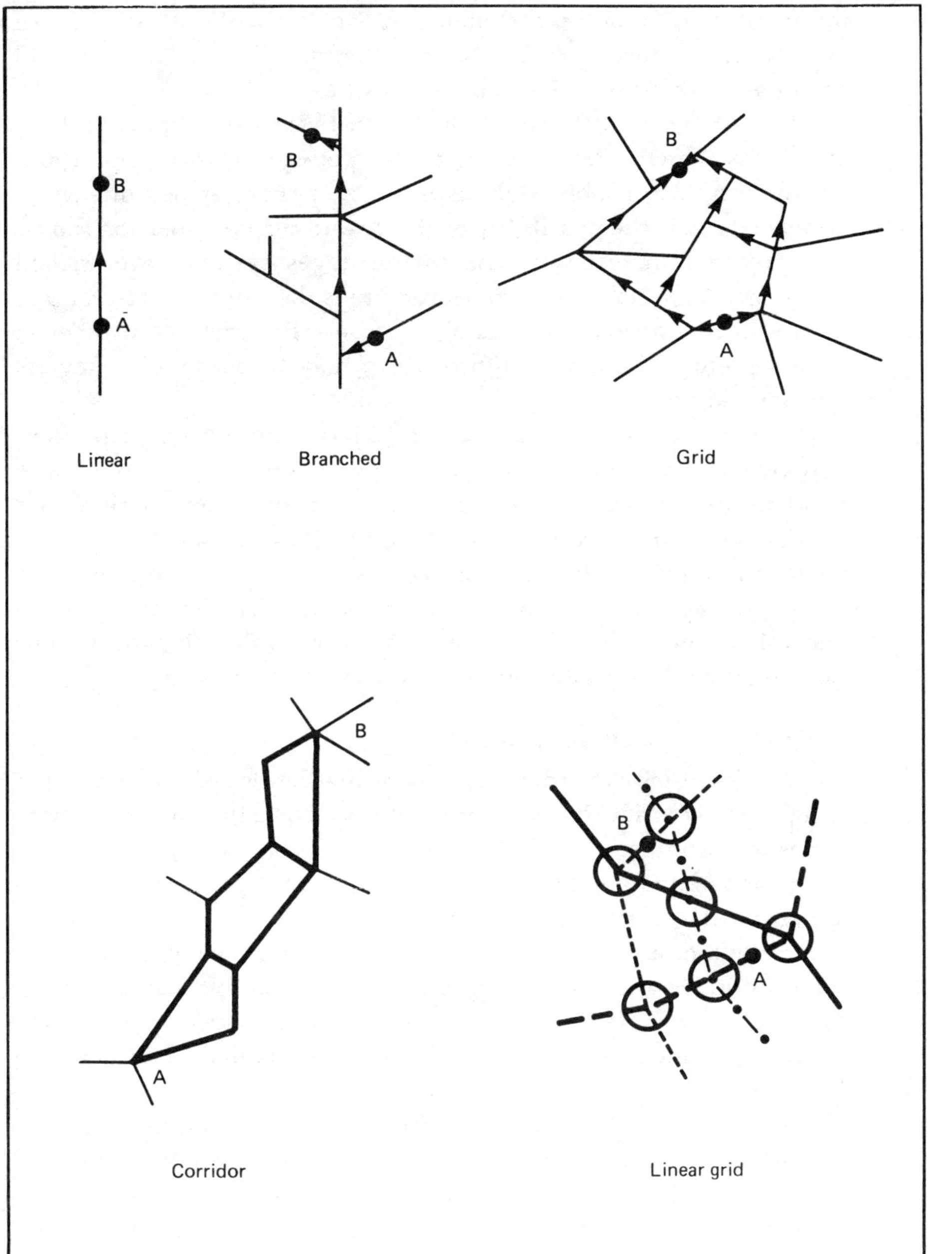

Fig. 4. Topological structure of railway networks

the shape of a star or the backbone of a fish, but without loops. These offer several routes, each one perfectly defined. This structure is still common on systems in developing countries.

Grid systems are differentiated from the first two by the existence of loop lines which afford several routes between starting point and destination. The problem of dispatch then changes in dimension, because there is the possibility of choice and the best solution can be found by combinatorial analysis. All the biggest systems have reached this stage. A special case of this structure is the 'corridor' or group of lines which run practically parallel and are linked to each other at certain points. The length of a corridor may be as much as several thousand kilometres.

There is also a combined linear–grid structure illustrated particularly by metropolitan subways. These are generally made up of independent lines, but passengers can change at connecting stations and thus enjoy all the possibilities of the grid networks; the best route to take is up to the individual passenger, according to the number and the length of connections. As for goods, the transfer of containers from one train to another in specialised yards, without shunting the wagons carrying them, is equally dependent on a linear–grid structure.

123 *Operational parameters*

Essentially four of these may be distinguished: the maximum weight per axle; speed; the number of tracks on a line; and the length of passing sidings.

The maximum weight per axle is a complicated parameter in which adhesion, the formation, stability and maintenance of the track, the types of materials used (weight per unit length), the nature of traffic, the train loads and their maximum length are all influential factors. This last factor affects the length of the track used in passing and shunting sidings, and so is a parameter of particular importance for goods traffic.

There are two notions of speed. First, the running speed, where the maximum value on a given line, sometimes called the 'potential speed', is fixed by geometrical parameters, by the type of signalling for passenger traffic, and by the characteristics of the rolling stock for goods traffic. Then there is the commercial speed, which is calculated from the whole rail journey from starting-point to destination, and which takes particular account of stops and dispatch methods.

The relationship between the two speeds, or speed efficiency, is one of the best criteria for determining the effectiveness of operation.

The number of tracks on a line introduces the idea of operational capacity. By its very nature, railway track is two-way, yet the majority of the lines on the world's railway systems are single-track.

The length of sidings defines the maximum capacity of the train. It is a parameter very costly and slow to modify. So for a new line, its fixing assumes great importance.

124 *Economic parameters*

These come into force at the level of management and sales, in particular in fixing commercial policies.

The essential parameter here is the economic policy of the country in which the railway is operating. In theory there are two kinds of policy, free and planned, but we shall see that in practice the options are not so clear-cut. Another significant economic parameter is the value of time. All transportation involves a length of time, and, where there are several types of transport offering a similar service, the choice depends at least in part on an appreciation of the value of time. It is obvious that this notion is largely subjective in the case of passenger transport, and is difficult to isolate from other factors, such as the timetable and comfort. It can more easily be worked out for the regular transportation of goods, as a function of stock rotation.

13 **Systems**

Before describing the modern railway by means of systems analysis, it is a good idea to recall several notions. The definition of a system given in the introduction – 'the sum of the elements in dynamic interaction, organised as a function of a goal' – contains three essential elements: the choice of an objective; cyclical functioning; and interaction or feedback.

The link between these elements is 'real time'. It is said that a system works in real time when every corrective decision (feedback oriented towards the objective) may be taken and executed in a very short time as compared to that corresponding to modifications of data (cyclical functioning). The unit of real time can only be defined relative to a fixed cycle.

131 *Management rhythms*

In the management of every enterprise, and in particular a railway system, we can see three fundamental cycles or rhythms.

Operational rhythm expresses the idea of continuous operation, as in a production line. In the case of the railway it is the running of trains.

Tactical rhythm has as its basis the idea of continuous administration. Its unit of real time may vary between the week, the month, or even the year, but this does not alter its importance.

Strategic rhythm has to do with the idea of planning. It applies to research, to investments, to staff recruitment, to commercial policies, etc. Its unit of real time is several years.

The organisation of goods traffic clearly illustrates the interdependence of these three rhythms. Studying the continuous running of the service – the operational rhythm – makes it possible to rectify and improve timetables and dispatch plans for the next period of service – the tactical rhythm. Long term developments, combined with a study of the market, lead to investments in materials and equipment and so on – the strategic rhythm.

132 *Real time on the railway*

Strategic and tactical units of real time make well-thought out decisions possible from an analysis of past information and possibly by using simulation techniques. The use of data processing, telexes, analogue computers, and methods of operational research and linear programming has transformed the range of possibilities in this area.

The unit of real operational time on the railway has, in fact, diminished in a spectacular way over the last twenty years. For more than a century, in effect, it was a minute in length because most operations were either largely or totally manual: switching points or signals, transmitting a message, drafting printed matter, etc. But with the increase in the speed of trains, intensive use of lines and the development of communication this is no longer the case. The unit of real operational time has of necessity become about ten seconds. This development has brought about a fundamental change. A man can no longer take all operational decisions alone; increasingly often these are prepared for him by automatic procedures which may continue, in a normal system, up to the command for the action to be executed.

Here the close link between 'systems' and 'cybernetics' becomes apparent. Cybernetics is actually both 'the science of control and communication in animals and machines' according to the classic definition by Norbert Wiener, and 'everything that enables organisation by means of computers and which makes automation possible' (Louis Armand). Without cybernetics, we could have neither the adaptability nor techniques of systems analysis.

133 *Internal subsystems*

The system making up a railway is obviously of very great complexity. So it is convenient to break it down into subsystems. These can be divided into two kinds:

(*a*) the functional subsystems one finds in every enterprise: production, sales, financial management, maintenance, etc.

(*b*) Geographical subsystems, which arise from the physical characteristics of the railway and from its own particular restrictions. The operation of a marshalling yard or of a line constitutes a geographical subsystem.

In the case of a specialised railway with totally integrated organisation, such as a mining line or metropolitan subway, these two kinds are not clearly distinguishable. But most often subsystems are distinct and strongly influence one another. An example of this is the choice of the criterion of priority in operating a line (the geographical subsystem), whether it should be passenger traffic or goods traffic (functional subsystems).

134 *External systems*

The railway, like other modes of transport, is only one of the instruments of the economy. It cannot therefore be studied in isolation, and it is useful to identify and analyse the various external systems the railway system influences and which influence it. Essentially they are the following:

(*a*) the users who are by definition associated with all production.

(*b*) the state, whose influence is enormous whatever the economic strategy of the country, and covers three main areas: transport policy, which affects commercial policy; financial management, and standards of safety and conditions.

(*c*) other forms of transport which provide systems similar to the railway system. There is a lot of feedback: technological developments modify the range of supply and relative costs, state intervention influences the distribution of markets. The idea of the 'transport chain' leads to the creation of intermodal subsystems.

(*d*) ecology, which may be defined as 'the science of the relation of the organism to the external world, i.e. in the global sense, the science of the conditions of existence' (Haeckel). For several decades, ecology has been playing a more and more important role in the transport sector. The principal relationships between the railway and ecology are the use of land, particularly in urban areas, pollution, noise, and site security.

135 *Railway macrosystems*

Lines or systems were initially isolated, but were soon linked up to make what we shall call railway macrosystems (RMS). Nowadays these play a major role in the development of traffic. Their operation has led to a degree of standardisation in all the operational areas, and to the establishment of decision or coordination bodies.

14 Plan of the book

Figure 5 is a schema of the different systems and their interactions, while figure 6 shows in greater detail the interactions at the level of the functional subsystems of a railway network. It also serves to clarify the plan I shall follow in the rest of this book.

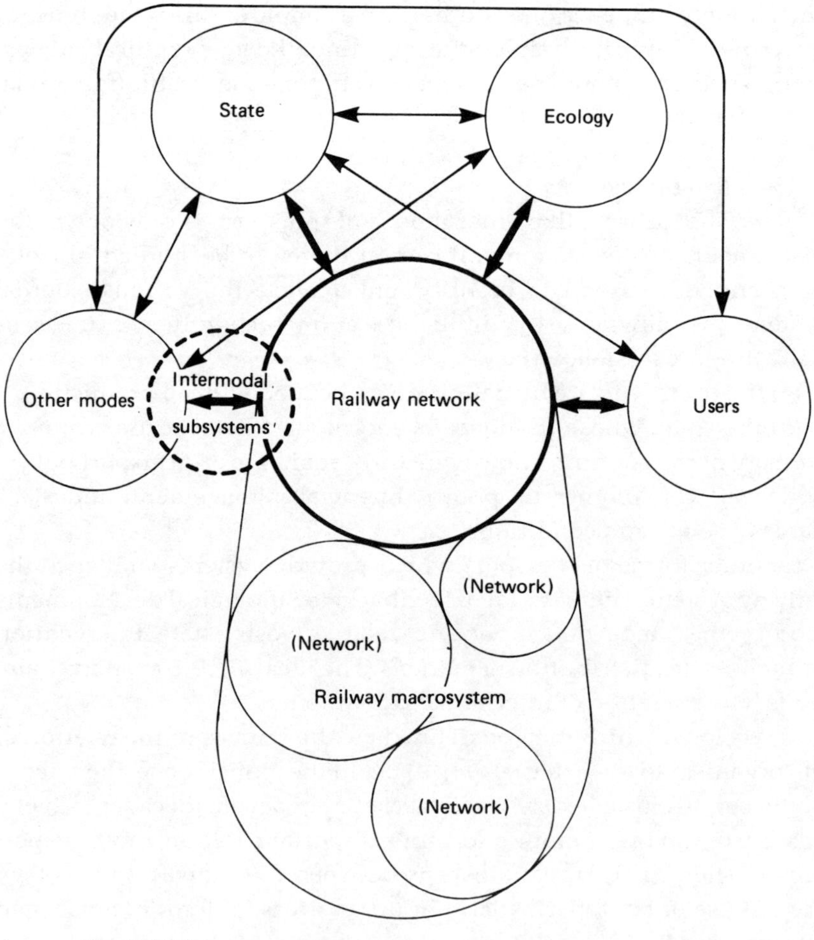

Fig. 5. Different systems and their interactions

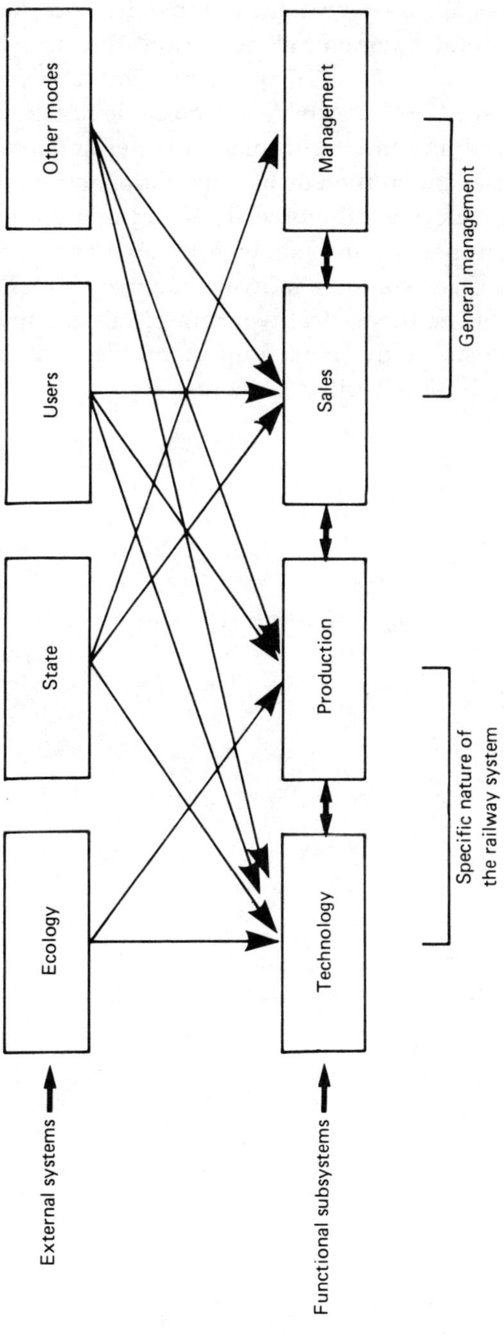

Fig. 6. Principal interactions at the level of functional subsystems

First, I shall briefly describe railway technology, about which many specialised works have been written, and then in more detail, railway production, which railwaymen call 'operation', and which is the sole truly original function of the railway. Having thus analysed the tools of the railway, a study of sales makes it possible to take up next the different types of supply and the problems of intermodal co-operation. A chapter on management then deals with national transport policies and their consequences for the objectives and operation of the railways; it will try in particular to answer the question 'who in reality holds the power of decision in a railway management?' Then follows an examination of the theoretical working of railway macrosystems and a brief analysis of the most important ones. A final chapter describes the international railway organisations.

2

Technology

Railway technology is not an autonomous applied science. It makes demands on the many disciplines making up the art of the engineer: civil engineering, metallurgy, electrotechnology, electronics, power systems, control systems and so on. A very rich literature, usually specialised and scattered, deals with the different aspects of modern railway technology. I have restricted myself in this chapter to those objectives constantly in the minds of engineers and decision makers, and to assessing the present situation. The structure of this chapter follows my systematic approach. After examining successively the infrastructure, the track, and vehicles (powered and hauled), I have regrouped them under the headings of the track–vehicle couple, traction and safety so as to consider together questions which the functional divisions of railway administration most often distribute among two or three departments. Such a method of presentation seems to me to bring out the increasingly multidisciplinary nature of the problems faced by railwaymen. To conclude the chapter, I shall comment on the main areas of research.

21 **Objectives**

211 *Response to demand*

The obviousness of this objective should not make us lose sight of the fact that it has only gradually become important. It was natural that the railway, born out of technology, should give priority to technology. This it did successfully during its first century of expansion and of virtual monopoly. Operation programmes then depended on constantly improving possibilities provided by technology, especially for railways which bought their equipment direct from manufacturers. Other railways however, among them the largest ones, soon felt the need to create their own research departments, so as to optimise supply and demand in parallel. Today, criteria of demand – economic, commercial and operational – must needs be the basis for technological developments.

212 *Maximising safety*

Every day about one tenth of the world's population uses the train, making journeys ranging from a few to several thousand kilometres. This simple reminder locates the problem of safety. From the beginning railwaymen and governments have placed safety at the head of their concerns. They have succeeded well enough too, because the railway is by far the safest mode of passenger transport. Of course there will always be accidents, sometimes catastrophic ones, but the coverage given such events by the media serves to underline how rarely they occur. Unlike other forms of transport, railway companies do not normally take out insurance for passengers or goods carried, nor for their equipment.

Technology plays an important role in the improvement of safety in two ways. The first is the strength of equipment, which is affected by progress in metallurgy and metallic construction. As with other modes of transport, no universal standards exist, for safety has evaded all statistical definition. The second is safeguarding traffic, which is affected by the development of electronics and control systems.

213 *Minimising costs*

Common to every enterprise, this objective is realised by railways in the following ways:

(*a*) reducing expenditure on staff, which as we shall see is still enormous, by automation. Because of the guiding system of railways, the various tools of cybernetics, such as electronics, data processing and control systems provide many possibilities in this respect, which are still only partly exploited.

(*b*) saving energy by installing highly efficient tractive systems.

(*c*) reducing the weight of rolling stock so as to increase the payload of trains. Such reductions, which may conflict with safety standards, depend on developments in metallurgy and plastics.

(*d*) adopting highly reliable techniques of value analysis and preventive detection of costs.

(*e*) standardising dimensions and specifications, making it possible to reduce the number of types of equipment. This is fundamental at the level of macrosystems, yet with the exception of the USSR and the US, railway equipment is rarely produced in runs larger than 1000 units for locomotives and passenger coaches, a figure which rivals aeronautical construction, and 10 000 units for goods wagons. These are derisory when compared with figures for car production.

214 *Avoiding obsolescence*

Railway operation is continuous whatever the weather or climatic conditions. Also, given all the safety requirements, railway equipment is of very solid construction. Its normal life-span of several decades is much longer than that of other forms of transport, such as aeroplanes, trucks and buses. This means there is little financial depreciation, but carries with it the risk of obsolescence, which may restrict the development of certain markets, such as comfort of coaches, specialisation of wagons and the facilities provided at stations.

All technical resources must be put to good use to ensure that fixed installations have lightweight structures which can be easily converted – with the exception of some passenger stations whose character as civic monuments is a matter of local town planning. The problem is more difficult with rolling stock; the use of modular construction and lightweight materials in fitting out vehicles must also be financially and commercially acceptable.

215 *Ecological considerations*

As far as land usage is concerned, the railway shows up in a favourable light in comparison to other land-based forms of transport, in particular roads and canals. An electrified railway with two tracks takes up only 10 m of land as against 35 m for a two lane-motorway. The maximum passenger returns per metre width is 30 times greater for rail – an undeniable advantage. However construction of new railway infrastructures is beginning to run into problems of site protection. Prior consultation with official bodies and representatives of the affected communities is essential.

The railway is also in a good position as far as pollution is concerned. Electric traction, exclusively used on urban and suburban networks, does not pollute. Diesel or gas turbine traction pollute far less than cars or trucks. Strict legislation in this this field is under study in several countries, although progress has yet to be made. On the other hand, the railway is noisy. It creates two types of noise: traffic noise, which originates from the rail–wheel couple, and operational noise, such as wagons crashing together in shunting yards or echoing announcements.

Some countries have begun to limit the maximum permitted levels of railway noise and without doubt most other countries will follow suit. The railway must prepare for this by technical means, and must avoid, for example, constructing marshalling yards or new lines which pass close to built-up areas.

It is therefore desirable to make railway constructions as attractive as possible. From the beginning engineers have built tunnels and viaducts in metal or stone which blend in well with the surrounding countryside. The same is true for stations. Collaboration with designers in choosing the shape and colour of rolling stock and in their interior decoration is also common practice.

22 Infrastructures

221 *Layout*

The order of priorities and the value of parameters which influence the layout vary with the objective fixed for each line. This depends on strategic management and will be examined in chapter 5.

For metropolitan subways and suburban lines the essential parameter is capacity. Urban geography dictates the layout, in particular the minimum radius of curvature, which may be below 100 m, and the maximum gradient, which may be as high as 4 %.

The layout of mining lines, which transport raw materials of little value sometimes over very long distances and often through desert regions, is influenced by the economics of operation. The most important parameter is maximum gradient in relation to load, so as to allow trains of very high tonnage. It is best not to exceed 5–6 mm/m, while the minimum radius of curvature is of secondary importance.

For lines open to all kinds of traffic the parameters defining the layout must above all be compatible with the existing system. Values of 150 km/h for maximum speed of traffic and 1000 m for the minimum radius are now good reference points; their adoption implies no special allowance for rolling stock or signals. The maximum gradient may not be optimised when cost increases due to the construction of bridges and tunnels are compared with reductions in cost due to shortening the route, and to savings in energy and time.

The layout of new lines constructed to mitigate overcrowding on existing routes reflects considerations similar to those which led to the construction of the motorway network parallel to the normal road network. The independence of the two layouts outside built-up areas and the reduction in the number of junctions make high speeds possible. The objective is more or less complete specialisation of the new line, which carries fast-moving traffic, either passengers or goods. Technical and economic studies show that the optimum speed is about 250–300 km/h, corresponding to a minimum radius of curvature of 3500–4000 m, the same as that of motorways on terrain of average irregularity. This may sometimes make twin infrastructures possible. If normal maximum gradients (10–15 mm/m) are observed, which is particularly important if the new line is to carry goods, costly layouts become necessary as soon as the terrain becomes a little uneven, and profitability is assured only if there is heavy traffic, for example the several hundred thousand passengers per day on the Japanese 'Shinkansen'. In order to change the height of the profitability threshold on such lines the SNCF adopted a maximum gradient of 35 mm/m for its Paris–South East line, which made it possible to reduce drastically the number of bridges and tunnels on moderately uneven terrain[1], and reduce the cost of construction to about 65 % of a two-lane motorway on the same terrain. Such a solution is technically possible only if engines with very high power-to-weight ratio are used and the number of driving wheels in the form of self-propelling trains are increased, which means that the line can be used only for passenger traffic. This totally new conception is therefore applicable only on 'corridors' which have heavy passenger traffic, i.e. tens of thousands of people per day.

222 *Studies and construction*

Studies of new lines have been expedited and made more precise by the modernisation of map-making techniques and the development of photographic surveying. Nowadays computers are used to optimise longitudinal profiles. The construction of the track formation of a railway depends, as it does for roads and canals, on civil engineering technology, which made considerable progress during and after the Second World War through mechanisation. However some new lines have given rise to specific problems, for example the stability of the track formation in extreme geoclimatic conditions such as permafrost on the Baikal–Amur line or moving sand dunes in Mauritania. There is also the case of tunnels whose length may exceed 50 km[1], as in Japan, and in plans for the Channel Tunnel and the St. Gotthard low-altitude tunnel. Here the contribution of seismology is becoming indispensable.

In general, work on infrastructure is not carried out by the railways themselves but, under their direction, by specialist firms. In some countries, public bodies are commissioned for construction of all transport infrastructures.

23 Track

The track consists of two rails kept a constant distance apart by sleepers. The sleepers also ensure an even distribution of forces on the track-bed through the intermediary ballast.

231 *Rails*

These are shaped like the bottom of an ice-skate. The quality of steel must meet many demands of safety and of use. Developments in metallurgy, notably the assimilation of oxygen in steel, make it possible to have rails whose tensile strength exceeds 50 daN/mm^2. The linear mass and vertical inertia of the rail broadly determine the maximum load per axle. Rails of 45–60 kg/m, the most widely used, can carry 20–22 tonnes per axle. However particularly in the US, rails may weigh up to 75 kg/m. Greater mass is an important factor in determining the useful lifespan of a rail because it reduces maintenance costs.

Rails have a unit length of about 20 m and are joined together by fish-plates. The joints are points of weakness on the track, and create harmful shocks to both rolling stock and loads. After 1950 came a major advance whereby rails were welded together over long distances. This technique necessitated a detailed study of the properties of expansion as it affected safety. These problems have now been overcome, provided there are strict checks after the rails have been laid and increased surveillance during extreme temperatures. The length of rail welded together in this way may be forty or fifty kilometres. The use of long welded rails has provided a new level of comfort, especially at high speeds, and has also reduced maintenance costs significantly.

232 *Sleepers*

These may be of any one of three materials: wood, steel or concrete (usually prestressed). The choice is essentially economic. The number of sleepers per kilometre varies between 1500 and 2000, according to axle load. Attempts have been made to replace sleepers with continuous slabs of either ordinary or prestressed concrete; at present this method is restricted to tunnels, viaducts and bridges.

Attaching rails to sleepers is a delicate operation, so much so that, as we shall see, the rails are normally laid slightly conically for reasons of stability. Recently, pliable fastenings which partially absorb vertical stress have been added to traditional spikes and sleeper screws.

233 *Ballast*

This is an elastic mattress between the sleepers and the track-bed. As well as distributing stress, it helps to stabilise the track and assists drainage independently of the distribution of stress. Ballast if often composed of several layers of various fragmented materials. Its thickness depends particularly on the axle load but is usually 15–35 cm.

234 *Points*

As we have seen, these are an inevitable consequence of guided transport. Crossing them does not impose any speed limit on the straight branch, and when installed on main lines they normally have inbuilt locking mechanisms. Higher speeds have led to very long points – 208.4 m for points on the SNCF Paris–South East line, which can be crossed at 220 km/h when switched.

24 Vehicles

Railway vehicles comprise tractive units, locomotives or rail-cars, and hauled vehicles, either coaches for the transport of passengers or wagons for the transport of goods.

241 *General conception*

A railway vehicle is made up of a body resting flexibly on two moving frames called bogies. Apart from the last steam locomotives, there are also some electric locomotives with several axles fixed rigidly to the body, but this form is being abandoned because it causes track fatigue and it is not suitable for high speeds. This is true of wagons too, and occasionally of coaches with two or three axles, which still make up a part of the rolling stock on some networks and macrosystems of Europe and the Sub-continent, but the trend is rapidly towards vehicles with bogies.

The objectives in the construction of rolling stock are:

(*a*) resistance to shock and fire;
(*b*) maximum use of loading gauge;
(*c*) lightweight materials;
(*d*) efficiency in the number of seats or tonnes per linear metre and an increase in the load/loss coefficient;
(*e*) passenger comfort and prevention of damage to goods;

(*f*) reduction in time for loading and unloading;
(*g*) standardisation either of a whole vehicle or its main components, such as bogies or auxiliary fittings.

The body is a shell cast in one piece, as is a car or an aeroplane. Welded metal construction has become standard. Developments in metallurgy have made possible the use of high-resistance steel and light aluminium alloys. Non-inflammable plastic materials are used more and more often for interior fittings.

Bogies usually have two, sometimes three and rarely four or more axles. Their design and method of connection to the body have greatly progressed as a result of research into stability, which will be described later. Some rail motor sets are articulated, with special bogies supporting either end of the adjoining bodies. Changing bogies or axles provides the simplest solution to the problem of different gauges. Sets of jacks allow the operation to be carried out simultaneously on several vehicles coupled together, in a very short time.

242 *Running gear*

Railway wheels are fixed on an axle body. Contact with the rail is maintained by a very hard tyre of steel, reinforced by a flange, and the rolling surface is slanted to make guiding easier. Use of wheels made in one piece, which eliminates the possibility of the tyres shifting, is spreading rapidly. The diameter of the wheels, which was sometimes more than 2 m on express steam locomotives, is now around about 1.25 m for diesel engines and 920 mm for hauled vehicles. It may be as small as 355 mm for some very low-slung vehicles. The free wheel used by automobiles has been the object of several experiments, but in fact only one of these has been put into operation. This is the TALGO system, in which the free wheel, connected to a short body of light construction, has made it possible to build articulated train sets with a very low centre of gravity and to obtain substantial increases in speeds on lines with winding routes.

Transmission of the load between the spring devices on the bogies and the axle joinal is effected by means of axle boxes. These are essential for safety but have long been a nightmare in operating a service because with existing technology (intermediary friction on a bearing of light composition, mechanically lubricated), there was frequent overheating, which if not discovered in time was likely to cause the journal to break, in turn causing a derailment. This technology has gradually given way to roller bearings or needle bearings, which have practically eliminated overheating. Otherwise, automatic overheating detectors may be installed along the track which are able to stop trains with suspected hot axle boxes.

The axles maintain a constant distance between the wheels, with normally about 2 cm play relative to the rails. Axles which have variable spacing and so able to change gauge automatically have been designed and experiments are now being made in Europe but exhaustive testing of fatigue and wear will be necessary to ensure the reliability of such devices. The TALGO free wheel may provide one solution.

243 *Coupling*

Coupling is a natural consequence of the convoy system. It performs two functions: it transmits the tractive force and absorbs shocks between vehicles.

Manual coupling separates these functions. A central coupling screw transmits the tractive force, while two lateral buffers, or occasionally a central buffer, absorb shocks. Modern manual couplings permit a maximum tractive force of 85 tonnes on flat ground. This consequently limits the tonnage of the trains, especially on slopes where the component of gravity comes into play, and rules out trains of more than 4500 tonnes gross. Manual coupling operations also carry a certain risk, made more significant by their frequency; several thousand such operations are carried out each day in a big shunting yard requiring a large unskilled work force. Manual coupling is one of the last survivors of the railway of 1830.

Automatic coupling, combining the two functions of traction and shock absorption, has been in operation for nearly a century. At first reserved for specialised equipment, such as rail motor sets, complete train loads, etc., it gradually spread to the level of networks and macrosystems, with the exception at present of Europe and the Subcontinent. Several designs exist, unfortunately incompatible with one another; the most recent also couple brake and electricity leads.

244 *Passenger coaches*

In a passenger coach the unladen weight is never less than 20–30 tonnes, while the payload rarely reaches 10 tonnes. Most of this weight (bogies, brakes, air conditioning, doors etc.) is independent of the load. It is therefore worthwhile to give the coach the greatest usable volume possible within the limits of the kinematic loading gauge.

The maximum length compatible with a loading gauge of 3 m is around 26 m. This has now been widely used for several decades for long-distance trains. Coaches on metropolitan subways are much shorter and not so wide because of the low radius of curvature.

The maximum use of the height allowed by the gauge is not particularly important unless it is possible to install two levels of seating along at least a part of the coach. Only the gauges of the US and USSR allow this without restricting comfort, and such designs, which make quite substantial savings, are being rapidly developed in the former. On long-distance coaches, with maximum capacity of no more than 100 passengers and used on trains which stop at stations for quite a few minutes, two doors which can let one person on or off at a time are sufficient, while only one door is enough for sleeping coaches. On the other hand, coaches on suburban and metropolitan lines need several large doors allowing 2–4 people to get on or off together because each stop is only for a minute or so. In this way it is possible to have a flow of 240 passengers per minute on a coach carrying 250 passengers.

The carriage–platform interface is important for the safety and comfort of passengers. A satisfactory type is the high platform, which comes up to the level of the coach floor. However a large number of railways have, at least on their main lines, low platforms only, raised slightly above the level of the rails, which means that access points have to be fitted with fixed or folding steps.

Comfort is a primary objective. Considerable progress has been made in the last 20 years and has led to continual increase in speed. The smoothness of running has been increased by a more scientific understanding of the track–coach couple. Now it is possible to travel in trains at speeds of up to 200 km/h in conditions at least as smooth as those of an aeroplane in a non-turbulent atmosphere. Soundproofing has been transformed by new insulating materials, very long welded rails on the track, streamlining the ends of coaches, and permanently closed windows fitted in conjunction with air conditioning. One can even sleep at 160 km/h. Electric heating has replaced steam and gives better control. Air conditioning is being more and more widely used, in spite of its high cost and the higher unladen weight of several tonnes that it entails. Its general use is becoming a necessity on a large number of railways because of improvements in living standards, and because of competition from aeroplanes and some long-distance buses which have it. Modern trains are also soundproofed.

In order to satisfy the different sectors of the passenger market the railway has gradually been led to create three basic designs for coaches: metropolitan and suburban; long-distance; and sleeping. Each design seeks to optimise a particular parameter.

Coaches (or rail motor sets) used on metropolitan subways or suburban lines are meant for short distances only and so give priority to capacity and access. The number of seating places is usually smaller than the number of standing places and capacity may be as high as 250 passengers, i.e. 15–20 per linear metre. Three or four large doors are fitted along each side.

Unlike these, long-distance coaches are principally designed for comfort and usually carry seated passengers only. They are either divided into compartments or left open; the compartment, inherited from the stage coach, was originally the rule but is quickly giving way to the open saloon. The width of the compartment and the number of bench seats vary according to comfort requirements. The extremes are compartments 1.75 m wide for 4–6 passengers and 2.5 m wide for 8–10 passengers. Thus a coach can provide anything from 30 to 100 places. The open saloon contains rows of one, two or three seats separated by a central corridor. The spacing of seats with inclining backs varies between 0.85 m and 1.20 m, so the seat ratio is similar to the compartment coach. Construction is more economical and maintenance rather easier, but on the other hand heavy luggage can only be placed at the ends of the coach and so cannot be as easily looked after by the passengers. An open saloon coach must have air conditioning and does not provide room in the corridors for people to relax.

Sleeping coaches are also designed with a view to comfort. The interiors are of two types. The 'couchette' is obtained by converting a long-distance compartment coach to make two or three tiers of bunks from one bench, giving sleeping compartments with 2–6 places. A couchette coach can thus be used both by night and by day. Capacity varies, according to comfort, between 30 and 60 sleeping passengers. Besides the couchette there is also the bed compartment. This is a real bedroom, with full beds either one on top of the other or on the same level, together with a washstand and sometimes a WC. The traditional bed compartment, with only one bench, washstand and with two or three tiers of beds, has a width varying between 1.25 m and 1.70 m giving a maximum of 36 beds per coach. Designers are trying to create compartments which have only one or two beds in a smaller volume by using several levels according to need. One completely full sleeping coach with large single beds carries only 0.5 passengers per linear metre and weighs more than 4 tonnes per passenger. Some long-distance trains with a largely tourist clientele offer a few very large compartments which are as luxurious as a hotel suite.

Sleeping coaches usually have a conductor and a mini-kitchen for preparing breakfast.

Apart from these there are specialist coaches offering supplementary services. The restaurant car, either traditional or self-service, may include room for 30–50 passengers and a kitchen; leisure cars may have bars, possibly shops, sitting rooms, a cinema, etc. There are also mail coaches, which allow sorting while the train is moving, and vans for transporting luggage, together with vehicles for carrying accompanied automobiles, usually on one or two levels, and preferably covered over. Nowadays coaches which perform various functions are being built: with both seating and sleeping places; seating places and a restaurant or bar; goods storage wagons with seating places; goods wagons with mail sorting areas; etc.

245 *Goods wagons*

Unlike coaches, the unladen weight of wagons is much less than their payload, and the primary objectives are to maximise the load/weight ratio, minimise length, facilitate handling, and reduce the risk of damage.

The maximum load/weight ratio is obtained by using welded wagon bodies and light metals or alloys. For uncovered mineral wagons the ratio approaches 4 : 1. The maximum load per axle permitted by the track is used, with possibly three axle bogies or double bogies on four axles.

Minimum length, important in splitting tracks and private sidings, is obtained by making full use of loading and clearance gauges. It is then possible to build covered wagons of 140 m^3 for every 20 m in length.

Handling is made easier by fitting large sliding doors in the sides, roofs that can be opened, and, for wagons used to transport loose or liquid materials, devices which discharge by gravity.

Risk of damage is reduced by fitting shock absorbers to certain types of wagon.

At first the railway had only four principal wagon types: covered; open; flat and tank wagons. The diversification of goods transported, the development of packaging and the influence of roads have led to an increase in the number of designs. This has resulted in specialisation that is not without drawbacks as far as efficiency of wagon use is concerned, which often return empty, and for standardisation, which is limited to the main components of the wagon. But it is necessary to respond to the market, and besides the use of homogeneous block trains permits excellent wagon turn around, as we shall see later.

The range of covered wagons has become larger with the use of wagons with wide doors, making the use of palettes easier; there are also wagons with roofs that open, refrigerated wagons (periodically loaded with ice), cold stores fitted with cooling machines, forced ventilation wagons for foodstuffs and animals, etc.

The open wagon, originally the most used, has now been widely abandoned in favour of open or closed hopper wagons which provide automatic handling of various goods such as coal, minerals, construction materials, cereals and various powders.

The flat wagon is still used a great deal for bulky goods such as metallurgical products, vehicles and agricultural machinery. It has seen two particularly rapid developments in the transport of large containers and cars. There may be two or three tiers according to loading gauge.

The variety of tank wagons, initially used almost exclusively for petroleum products, has considerably widened with the transportation of chemical products. Differences in composition and of hazards have meant that in practice there is one type of wagon for each product. The food industry also uses tank wagons.

Finally there are a few very special types: crucible wagons, for transporting liquid metal between iron and steel works over distances of up to 100 km; wagons for the transport of exceptional loads, up to 400 tonnes, which are composed of two detachable chassis and need up to 32 axles; and wagons used for the transport of nuclear waste.

25 The track–vehicle couple

251 *General*

The implementation of adhesion and rail–wheel guiding introduces a mechanical subsystem, the track–vehicle couple. Having described the static structure of its two components, we shall now examine their interaction under operational conditions.

The track is composed of elements which have different degrees of flexibility, and the vehicles consist of wagon bodies resting flexibly on bogies. All these elements are subjected to forces which are variable both in direction and intensity. Further, the track and the vehicles have faults of a necessarily uncertain or erratic nature, which create dynamic loads that are themselves uncertain or erratic. These phenomena were scientifically analysed only recently, in response to the demand for increased speed, greater load per axle and improvement in comfort. Research carried out in this area by a few European and Japanese railways has overturned a large number of previously accepted ideas and their results have made an essential contribution to the shape of the modern railway.

The discussion here will be limited to a summary analysis of the problems of stability on straight track, negotiating curves, the load per axle, and finally the noise caused by rolling wheels.

252 *Stability on straight track*

Vehicles exert three types of stress on the track. The first, longitudinal, due to acceleration and braking, is of minor importance and does not pose any particular problems.

The second, vertical stress, is due to the load borne by the wheels. Experiments show that track resistance to these stresses is usually resilient, so that the problem here is a matter of fatigue. The use of heavy section rails and heavy sleepers, especially concrete ones, and the reduction of non-suspended weight in the design of vehicles can provide solutions.

Finally there is transversal stress, essentially due to hunting of the bogies on the vehicle. These stresses are destructive, for even when they are of moderate size they may lead to permanent deformation of the track. Transverse stability of the vehicles is thus the essential objective and in order to achieve it very detailed analyses of rail–wheel guiding have been carried out.

Detailed study has shown that under normal conditions of running on straight track, the wheel flange does not actually play any part in guiding. Guiding results only from the play of the forces of contact between the surfaces of the rails and those of the diconical wheels. The flange comes into play only in special circumstances: negotiating curves and switches, after wear and tear, etc. Rail–wheel contact actually occurs, under stable conditions, within a rolling band about 1 cm wide, only half the normal meeting of the wheel and the track. The choice of angle of the diconical wheel rim is thus a matter of extreme importance so as to avoid either too much feedback and so increased hunting, or permanent pressure on one of the flanges. On the vehicles, the stress between the body and the bogies is usually moderate and it is possible to avoid vibrations, either dangerous or simply uncomfortable, between the bogies and the wagon bodies themselves. Problems of stability thus in the end come down to the transverse dynamics of the bogies.

The self-correction of hunting described above can only be maintained within certain limits. The study of the transverse stress of bogies brings into play a model with numerous parameters (hunting, twisting, rolling, etc.) which when resolved yield a 'critical speed' above which the hunting movement is self-perpetuating and carries the risk of permanent deformation of the track. Experiments, though difficult to carry out, have confirmed the theoretical calculations suggesting that the critical speed be made as high as possible and in all cases significantly higher than the normal traffic speed. This research has issued in a number of recommendations, which only become imperative if speeds are to exceed about 150 km/h: making the track components of rails and sleepers heavier; restricting the load per axle; reducing the conicity of the bands from 1/20, the value which was generally accepted before, to about 1/40, together with an inclination of 1/20 of the rails on the sleepers; lengthening the wheelbase of the bogies; reducing the non-suspended weight on the bogies.

This last point has particular application in a new conception of the proper positioning of traction motors, which are placed under the wagon bodies, and no longer incorporated into the bogies, and consequently in the development of transmissions between motor and axle which are compatible with this kind of construction. The final result was the development around 1970 of rail motor sets capable of travelling at 300 km/h on traditional tracks with as much safety and comfort as conventional trains running at 150 km/h.

The rigidity of chassis also poses some problems, especially on curves of small radius, because the two axles cannot both be perpendicular to the rails at once; the result is extraneous stress, wear on rails and flanges and even the risk of derailment. A few systems are experimenting with 'radial bogies', which allow each axle slight displacements, though not independently of each other, and so provide variable geometry.

253 *Curves*

Centrifugal force comes into play on curves. In order to keep the resultant of this force and of the weight of vehicles perpendicular to the plane of motion, a difference in level is created between the two rails. This is called superelevation. For a given radius and superelevation the 'equilibrium speed' can be obtained:

$$v = \sqrt{\frac{rEg}{G}}$$

where r = radius, E = superelevation, g = acceleration of gravity and G = gauge of the track.

However on the great majority of lines, trains run at very different speeds (50–80 km/h for heavy goods trains, 150–200 km/h for express passenger trains). For a number of trains therefore the superelevation is either too much or not enough. Too much superelevation wears the inner rail and increases friction on wheel flanges, thus increasing the tractive effort required. Not enough creates transverse stress which may be detrimental both to safety, because it increases the risk of derailment and track displacement, and to comfort. Finally the transition between straight and curved sections of track poses the problem of the ratio of variation of superelevation, which is difficult to solve on winding routes.

Railways which carry mostly express passenger traffic usually limit superelevation to 150–160 mm. For a curve of 1000 m the equilibrium speed with standard gauge is thus 100–120 km/h, which means that the fastest trains take the curves with superelevation a little too low. Among the solutions to compensate for this that have been considered, and put into practice on a small number of railways, one is to give the vehicle bodies an extra inclination upon entering the curve. Different mechanisms have been used, such as tilting and recall systems. However these are very difficult and expensive to make, so their development will doubtless be restricted to certain special cases, where the value of the time gained justifies their use. British Rail is currently experimenting with the Advanced Passenger Train (APT), which travels at high speeds using a tilting mechanism. The problem does not occur on new lines which have been built for high speeds, for it is taken into account in fixing the design characteristics of the line.

254 *Maximum axle load*

As noted in chapter 1 this problem mainly concerns goods traffic. Increasing maximum load improves the efficiency of the equipment but it also increases rail fatigue considerably, particularly at joints. It is therefore not possible to exceed a certain maximum speed if one wants to ensure that passenger traffic will continue in acceptable conditions of comfort and safety.

The value of the maximum load is significantly affected by the weight of the rails, the number of sleepers per kilometre and the thickness of the ballast. Most networks fix this maximum mass at 20–22 tonnes using rails of 50–70 kg per metre and 1800–2000 sleepers per kilometre. In the US, where loads exceed 30 tonnes per axle, the running of passenger trains above 100 km/h remains tolerable only if maintenance is very careful and expensive, and rolling stock very robust. Track in the USSR, which carries a much greater amount of both goods and passenger traffic, is on the other hand restricted to 22 tonnes per axle, with greater efficiency of rolling stock obtained by using bogies with three or four axles. This solution is much better for comfort and maintenance.

255 *Running noise*

This is an inevitable result of the steel-on-steel contact of the rail–wheel couple and creates a double nuisance to passengers and the environment. Since it is not possible to eliminate noise, any effort to lessen it must keep inside economically acceptable limits. The level of internal noise may be reduced by the use of soundproof materials on the floor of coaches; a threshold of 50–60 dB(A) is possible and seems to be sufficient, especially for night journeys. The level of external noise can be reduced only by the installation of screens, either fixed along the track (walls or cuttings) or mobile on the vehicles themselves (skirting on the bogies). Screen walls have been specially developed to cross urban areas on the Shinkansen line; the noise level may be reduced to around 85 dB(A). There are certain dimensional restraints which prevent the bogies being completely covered; this solution is still in the experimental stage. The influence of some track components, for example material used for sleepers, welded rails, etc. is clear but not understood very well at present.

Some mechanical noises (motors, ventilators, compressors, etc.) are drowned by running noise only at speeds over 100 km/h. They cause a nuisance at stops, particularly at stations. The only technique for diminishing them is to hood the noise sources, especially on shunting engines.

26 **Traction**

The railway was born and developed for over a century using steam locomotives fuelled by coal. However, though steam traction made it possible for trains to reach 100 km/h by 1860 and 200 km/h by 1935, and in spite of constant improvements made up to the Second World War, it has now lived out its days. Its energy efficiency is too low, it has harmful environmental effects which have become intolerable, and working conditions for the driving crew are terrible. The steam engine survives only on a few railways, notably in China, where coal, fuel oil, and even wood are cheap and economics have not yet made its replacement possible. We should pay richly deserved homage to the steam locomotive, for it also popularised the distinctive image of the railway in literature and the cinema.

Since the end of the nineteenth century, when new techniques of energy production and conversion were developed, the railway quickly assimilated these new possibilities to arrive at the present situation where various forms of energy are at its disposal: electric traction, which obtains its power from an external source, and self-propulsion, which produces its own power from motors. Of the latter, two types are currently in use: the diesel engine and the gas turbine.

Before describing the essential elements and comparing their performance, we shall look at the laws of mechanics governing convoy traffic.

261 *Resistance to motion*

Moving convoys have to overcome several types of resistance: resistance to horizontal linear motion; the positive or negative components of gravity on inclines; resistance due to curvature of the track; and resistance to acceleration.

Resistance to motion in a horizontal straight line is dependent on several parameters: speed, properties of the equipment, and especially the length of the convoy. Experimentally, it can be expressed as a parabolic function of speed:

$$R = A + Bv + Cv^2.$$

A here represents axle friction and resistance to rail–wheel movement on level track; it is proportional to the load per axle. B depends on several more or less clearly defined factors, in particular the wind and twisting motion; it is generally proportional to the load per axle and close to 0.01 daN per tonne. C is the effect of air resistance and is greatly affected by the factors determining the aerodynamical shape of the vehicles: frontal pressure, sloping of the rear of the convoy, streamlining and the space between vehicles, open windows, etc.

In addition, resistance to motion is influenced by the state of the track and higher in tunnels than in the open air.

By way of example, the aggregate resistance to the motion of a passenger train of 800 tonnes drawn by an 80 tonne locomotive is about 5500 daN at 140 km/h, and that of a goods train of 2000 tonnes drawn by a 120 tonne locomotive reaches 6000 daN at 80 km/h; but for a rail motor set of 400 tonnes which has been specially streamlined, it does not exceed 5000 daN at 250 km/h.

Resistance due to gravity is proportional to the convoy weight and to the sine of angle of inclination. Since the gradient is always low, the sine can be replaced by the angle, so that for a mass M

$$F \text{ (component parallel to the track)} = Mgi,$$

where g is the acceleration of gravity and i the angle of inclination. Thus an incline of 10 mm/m (1 %) requires an effort of 9.81 daN/t. Looking at the previous example again, if we take passenger trains of 800 tonnes travelling at 140 km/h, a 0.6 % slope doubles the resistance to motion of the convoy on level track. This confirms the topographical restrictions of the railway.

Resistance due to curvature of the track arises from the friction between wheel flanges and the lateral faces of the railhead, and especially from the wheels sliding on the rails as a result of the axles of a bogie or of a vehicle with a rigid wheelbase being forced to remain parallel. Curves increase the angle of inclination i by a value of k/r (r = radius of curvature). The coefficient k is generally estimated to be between 500 and 1000 according to the gauge (for r in metres).

Resistance to acceleration is given by the expression $\alpha M\gamma$, α being a coefficient of increase which varies with the equipment used, and which depends on the inertia of revolving masses; it is close to 1 for hauled vehicles and slightly higher for powered vehicles. For modern trains hauled by locomotives, the coefficient of inertia of all moving masses is itself slightly greater than 1. Finally, experiment shows that an acceleration of 1 cm/s^2 requires a tractive effort of 1 daN/t, or the practical equivalent of a 1 % incline.

262 *Convoy dynamics*

If we ignore extraneous movements which interfere only slightly with propulsion the system of forces applied to an engine working by adhesion includes the following: its weight, vertical reaction of the rails at the points of wheel–rail contact; convoy resistance applied by the intermediary of coupling; and the torque applied to the driving wheels to make them rotate.

At the commencement of this rotation (pseudo-sliding), there is a reaction of adhesion on the rail, which is the stress at the rim of the wheels. Composing all these reduces the system of forces to (*a*) a force *F* applied to the axle, and by the axle to the bogie; it is equal to the stress at the wheel rim;

(*b*) a couple of forces which make the wheel turn; this is equal to the product of *F* and the radius of the wheel R, i.e. FR;

(*c*) Convoy resistance; this is transmitted, with modern traction, to the body of the traction motor, which transmits it to the bogie. If this force is applied to the bogie above the axles, it combines with the force propelling the axles to produce a couple of forces which causes the bogie to swing; this phenomenon of nose-lift is avoided in modern locomotives by applying the link between the body and the bogie at the level of the axles.

Finally the propulsion of a convoy can be described in mechanical terms by considering the turning of a wheel subjected to a couple C equal to FR. The force F sets the convoy in motion, starts an acceleration then balances its resistance when the normal running speed is reached. It is very high at low or zero speed, and in theory its only limit is adhesion.

The strength of the convoy's propulsion is FV (V is the running velocity). It is zero on starting, but can increase to the limit of the engine's power. The versatility of an engine and its range of use depend on the possibility of developing this power over as wide a range of speeds as possible. The optimal design would meet the following objectives:

(*a*) exert a driving force at the wheel-rim, limited by adhesion, up to the speed corresponding to the maximum power of the engine;

(*b*) provide decreasing or at least constant power up to the speed limit;

(*c*) highly efficient over the whole range of operation.

Another very important characteristic should be added to all the others. The nominal power of the engine is in theory that which it can maintain indefinitely. But greater power, which can be maintained for short periods, such as a few minutes or maybe an hour, may also be considered. The capacity on starting or in special circumstances to deliver and maintain such power bursts is very valuable and well worth taking into consideration in the design of engines.

Lastly it should be mentioned that the increase in speeds has led to new thoughts on the use of the kinetic energy accumulated by a convoy and recoverable when the speed is lessened. The SNCF has made great use of this on its new Paris–South East line: a train of 6200 kW going up a slope of 35 mm/m at 260 km/h can travel 122 m before its speed slows to 220 km/h; double the power would have had to be installed to obtain the same result from an initial speed of 220 km/h.

263 *Electric traction*

A traction circuit consists of a source, a contact wire, traction motor and a current-return wire. The source is a substation, a relay between the primary supply network and the control circuit of the motor. The contact wire is very often a suspended wire or catenary; but contact may be achieved by means of a third rail, particularly on metropolitan subways and suburban lines which have a low voltage (600–800 V direct current), though this does not allow speeds of more than 160 km/h (BR). The return current is usually carried by the rail.

The fundamental choice of railway electrification depends on the type of current delivered by the substations. At the moment there are three types in use, whose advantages and disadvantages vary according to the state of technology, the type of traffic, the nature of the primary network, the effect on the environment, and finally compatibility with previous electrification systems.

In the direct current system, the current is carried on the contact wire at the same voltage used by the motors; in practice this is a multiple of 750 V with an upper limit of 6000 V. These figures are low for transport voltage and require high current strengths in the contact wires, calling for large-section overhead wires and substations set close together (up to 15 km for 1500 V). Such a system requires rather heavy installations.

In systems using single-phase alternating current at a special frequency, the voltage convertor in the motor is a transformer, making it possible to choose high voltages for the traction network (10–15000 V). The current strength necessary is low, the overhead wires are appreciably lighter and the substations, which are simple transformers, can be much further apart (50–60 km). But to solve the problems of commutation associated with direct current motors, a much lower frequency than normal, in practice one third of the industrial frequency, has to be adopted (16 2/3 Hz). This makes necessary either a primary supply network with such a frequency which is separate from the general network, or frequency conversion stations.

In systems using single-phase alternating current at industrial frequency, the voltage conversion in the engine is the same as for the previous system, except that the traction network is connected directly to the general primary network at 50 or 60 Hz. This system combines the advantages of the two previous ones, notably the light installations (overhead wires and substations). It was developed by Louis Armand around 1955 after a great deal of experimentation and research on locomotives, first with direct current engines, then with direct monophase currents and monophase–triphase groups and leading finally to locomotives with static current rectifiers, which were adapted and perfected (thyristors). The single-phase current of the overhead wires, whose voltage is usually 25 kV, is converted into a rectified current feeding the traction engines, which may then be similar to the traditional direct current engines but without constraints such as a starting rheostat or successive coupling of engines because the transformer supplies the rectifiers and the engines at a gradually variable voltage. On high-speed lines such as the JNR and SNCF, the overhead wires are supplied by autotransformers at 50 kV. Finally, some US railways and South African Railways are experimenting with a voltage of 50 kV on mining lines which need a lot of power.

Single-phase traction at industrial frequency quickly came into use throughout the world on lines electrified after 1960, except when it was not compatible with existing systems. The adoption of multicurrent locomotives has also allowed several networks to develop two systems simultaneously, the choice being made line by line according to local conditions. At present 140 000 km of track in the world have been electrified, distributed approximately as follows:

(*a*) 50 % direct current (suburban lines, western Europe, USSR, the Maghreb, South Africa);
(*b*) 30 % single phase at industrial frequency (throughout the world);
(*c*) 20 % single phase at a frequency of $16\frac{2}{3}$ Hz (Central Europe and Scandinavia, AMTRAK).

The considerable progress made in electric traction on the railways since 1950 has affected all the elements of the traction circuit. The average efficiency at full power of the substations is at present 98 %; their installations have been simplified and no longer need a resident staff, being controlled automatically from a few central posts in communication with traffic control centres. The section size of contact wires made of 'copper equivalent' has changed from 400 mm^2 for 1.5 kV direct current to 150 mm^2 for single phase current. Their suspension has been improved and the overhead wire–pantograph couple is now operable at 300 km/h. Traction motors have benefited from all the improvements in electronic power (thyristors, choppers), and recently the non-synchronous three-phase engine has made a significant appearance on the DB and CFF. Finally, the disturbances due to current return have been reduced by the use of various techniques.

Modern electric locomotives, whose power in continuous operation may reach 6000 kW, can now attain specific power output of the order of 50 kW per tonne. What is more, they can take considerable temporary overloading. This explains the undeniable superiority of electric traction on lines which have heavy or dense traffic, difficult gradients or very high speeds.

264 *Diesel traction*

Diesel traction is an autonomous system and depends on just two components: the engine and its transmission.

The characteristics of the diesel engine offer some advantages to the railway, but these are tempered by several constraints. A diesel engine offers good thermal efficiency, but low power-to-weight ratio. There is a close relationship between the functional and dimensional parameters (gauge and load per axle). Ultimately it is the cubic capacity which determines the space occupied by the engine; normally the largest have 16 cylinders arranged in a vee-shape. Two- and four-stroke engines are used; the first is very robust and gives a power gain of around 50 %, while the second avoids the difficulties of cylinder sweep. Improvements in the working of a diesel engine have been pursued in two main directions: increased power per litre by means of supercharging, and higher revolution speed, which makes possible an increase in the installed capacity. However there are no systematic developments in the range of 750–1500 revolutions per minute. Nowadays locomotives of 3000–4000 kW are constructed, having a power-to-weight ratio of 0.20–0.25 kW per kg and a volumetric ratio of 16–20 kW per litre. Traction in multiple units is very widespread, with energy-saving devices on sections with easy gradients.

A diesel engine cannot start under load and develop a usable torque until it is close to its cruising speed. It must have a transmission control system, which makes it possible for the engine to start and to run at low speeds, to choose the driving power and speed for economic running of the engine, and to ensure dynamic braking power. Mechanical and hydromechanical transmission are often used for shunting engines and rail motor sets of low or average power. Electric and hydrodynamic transmissions are both used for main-line engines. In addition, for the passenger service the diesel locomotive must be fitted with a heavy and bulky heating system, unless the rolling stock has its own air conditioning.

There are two main systems using diesel locomotives. The first is the American system where the objective was, and still is, to replace steam completely by diesel traction. A generous loading gauge and a high axle-load have led to the development of massive two-stroke engines.

The second is the European system, where dieselisation was only really developed after 1945, when some main lines had already been electrified. The limits imposed by loading gauge and the axle-load are also much tighter. In the beginning, shunting locomotives and rail motor sets were built the most often. On these systems, the comparison of possible power-to-weight ratios usually means that diesel traction plays a complementary role to electric traction. The choice between diesel and electrification varies according to the volume of traffic, the percentage of high speed passenger traffic, the price of oil and the amount of energy the country can supply by itself. Present developments have lowered the profitability threshold of electrification, and the US is beginning to consider electrification on a few main routes.

Developing countries have naturally opted for diesel traction; the volume of their traffic justifies electrification only on a few lines.

265 *Gas turbine traction*

The gas turbine, which is a rotating engine, eliminates a few of the mechanical restrictions of the diesel engine and has lighter construction. Thus it gives a rather higher power-to-weight ratio. The aeronautical industry has developed it to a high level of reliability. From 1940, several American and European railways tried them in powerful units using low efficiency turbines, but this approach was then abandoned.

Experiments were taken up again in 1965 on some of the same railways, with a view to obtaining high speeds on non-electrified lines, but this time using free turbines or turbo-engines, with the gas generator and the working turbine mechanically independent of each other. The torque increases as the speed diminishes so the engine can be used over a wide range of speeds. These turbo-engines have relatively low power (around 1000 kW); they are light and do not take up space, being directly derived from those used in helicopters though they burn diesel oil instead of kerosene (which is more costly and inflammable). In this way a power-to-weight ratio of about 2.5 kW/kg and a volumetric ratio of 100 kW/litre are obtained. Fuel consumption and maintenance costs, on the other hand, are greater than for diesel traction. The sole practical application is to rail motor sets: these can reach speeds up to 300 km/h, with an installed capacity of 4400 kW and electric transmission on all the axles. The present rises in oil prices do not favour the development of the gas turbine.

266 *Evaluation of energy use*

For about 30 years, energy use on the railway has greatly improved as a result of electrification and dieselisation. This can be analysed under three aspects:

(*a*) the energetic efficiency of the engines;

(*b*) low resistance to motion and low fuel consumption which results from it;

(*c*) the particular suitability of railway traction to various sources of energy, notably to electrical energy.

Energetic efficiency is the relationship between the energy supplied to the draw bar of an engine and the energy potentially available. Steam traction has a particularly low efficiency: a maximum of 8 % on the last generation of engines.

In electric traction, there are two efficiencies to be considered. The first is that between the generating station and the high tension terminal at the substations. This is about 0.95 (the average efficiency of a thermal power station is 0.35 while a hydro-electric station may reach 0.9). The second is between the substation and the locomotive draw bar; this nowadays reaches an annual average of 0.72 for single-phase current and 0.69 for direct current. The global annual average efficiency is thus around 0.25 if the primary energy supply is thermal in origin and 0.50 if it is hydro-electric.

In thermal traction, the efficiency of the diesel engine is around 0.38–0.40; if auxiliary engines and transmission are included, the average annual efficiency at the draw bar of a diesel locomotive is more than 0.22. For rail motor sets we can only compare the actual efficiency of the whole train at the best operating rate for the engines; this is 0.34 for a supercharged diesel engine and 0.20 for a turbo-train. In order to compare these figures with previous ones we must make a correction of about 0.9, corresponding to the locomotive's resistance to movement.

Modern railway engines, whether electric, diesel or turbine, have much greater energy efficiencies than steam traction and the railway is as efficient as road transport (about 0.22 for a car and 0.3 for a truck). Apart from this, modern railway engines consume considerably less energy than other modes of transport. This advantage, as we saw in chapter 1, essentially results from the contact of the steel rail on steel wheel, and from the comparatively low value of mechanical and aerodynamical resistance to motion. The forces necessary to transport a unit of traffic according to a fixed timetable, are, under these conditions, much lower than for road and air traffic. For example, if we calculated the resistance to motion of a tourist coach and of a passenger train, the power needed to overcome resistance to reach a speed of 100 km/h are, respectively, 4.5 kW for the automobile and 0.9 kW for the train, or a ratio of 5 : 1 in favour of the train. This ratio is slightly higher if we consider the distance of 450 km travelled either by a high-speed train like the SNCF's TGV or an aeroplane like the Airbus. Also, if we compare the resistance to motion for a speed of 80 km/h of a road vehicle of 35 tonnes carrying 22.5 tonnes of goods, and that of a train of 1000 tonnes transporting 630 tonnes net, the required power rises, per tonne transported, to 5 kW for the road vehicle and to 1.1 kW for the train, or a ratio of more than 4 : 1 in the railway's favour.

Such calculations may be wide-ranging, bringing into consideration the different coefficients of equivalence between forms of energy and numerous other factors, such as the load factor of the vehicles, the contour of the track, the speed and composition of convoys, the number of stops and starts, the time of year, etc. But the results always favour the railway, except in the special case of the motor coach, whose fuel consumption per passenger-kilometre is about the same as a railcar of average power without trailer.

Finally, energy sources must be considered when making an overall evaluation of energy use in railway traction. In this respect, thermal traction has the disadvantage of consuming only one fuel element, which is similar to domestic fuel, the main product of oil refining. With electric traction on the other hand, the railway can use a whole range of primary energy sources: coal, oil, natural gas, hydro-electric, geothermal and nuclear power. This last source, which will be developed in the near future, can only be used by land-based transport in the form of electricity and can therefore only be used by the railway. At present, when most countries are trying to find as many different energy sources as possible, this becomes a positive factor for the future of the railway. For the balance of payments of countries which do not have their own oil supply, the consequences are not inconsiderable.

27 Safety

Safety is the major preoccupation of the technical running of the service. It assumes three aspects, which are partially connected, according to whether it concerns users, staff, or third parties. The safety of a guided system has certain characteristics. The person driving a train cannot, as can a car driver or a ship or aeroplane pilot, change its course if there is danger. Two requirements result from this. The first is to give the train driver traffic commands: this is the part played by the signalling system. The second is to provide the material means to carry out these commands; this is the role of braking. The signalling–braking couple constitutes a functional subsystem, for which cybernetics has supplied new tools.

271 *Signalling*

Signalling must make it possible for the train driver to stop before meeting any obstacle, yet at the same time give minimal reduction in the use of the line. This excludes the possibility of travelling simply with a permanent view of the track ahead, and so requires the use of a signalling code. Braking cannot be immediate, so the code must give prior announcement when to start braking, so that the train will stop up line of the obstacle. The universal railway code, subsequently adopted by other modes of transport, therefore includes three signals: line clear (green); warning (yellow); stop (red). The distance between the warning signal and the stop signal corresponds to the braking distance plus a safety margin.

More precisely, the signalling system must avoid three possible causes of accidents: a collision at track junctions, for which there is the stop signal system; when a train catching up with one in front of it on the same line, for which there is the distance-spacing signalling system; and two trains travelling in opposite directions on the same track, for which there are face to face signalling systems on a single track.

A stop signal is directly connected to the position of the points to be protected, interlocking the setting of signals and points. The binary nature of point systems makes it readily possible to construct mechanical or electrical interlocking devices. Interlocking on the approach track prevents the signals from being set too late, and ensures that before setting a signal which clears access to another line, all points are in the correct position and will stay there. This is the particular role of locking and position control mechanisms.

All modern signals lie on a 'track circuit'. The line is divided into sections electrically isolated from one another; the passage of a wheel establishes an electric circuit between the rails which activates the relays and causes the signals and points to switch. The track circuit can be controlled by a very simple cybernetic microsystem, which has transformed reliability and the unit or real time in signalling.

The switching of signals and points is co-ordinated in signal boxes, which are one of the characteristic features of the railway. Instead of manual levers, with cable or hydraulic transmission devices, switches controlled electrically and partly automatically are being used more and more, thanks to track circuits. The area covered by a modern signal box may include several hundred routes and cover line sections of more than a hundred kilometres. Visual control boards give continuous information about the setting of the points and signals and which tracks are occupied; sometimes the type of train can be identified. The recording of several successive train routes, which are automatically cleared, is also in wide use in larger signal boxes. The two functions of safety and production control therefore have a tendency to merge into one.

The safety of traffic off the main lines, for example inside marshalling yards or depots, is often controlled by simplified signal codes. These do not have warning signals because the driver can always see where he is going.

The signalling for the spacing of trains is affected by dividing the lines into block sections which take only one train at a time. Each block section is demarcated by two stop signals, preceded by their warning signals. The opening of the entry signal on a section may be manual or automatic. Manual opening may be authorised by a simple exchange of messages; this is the 'block telegraphic or telephone' dispatch used on lines with light traffic only. Safety is increased by using interlocking devices which prevent premature manoeuvres; this is the 'interlocked block', used on lines of average and heavy traffic. Automatic opening is executed by the 'automatic block', which amounts to a cybernetic system; it is the progress of the train itself which manoeuvres the signals situated down the line, through the use of track circuits. Automatic block sections are usually the same length as the stopping distance and the same signal can display the three instructions, clear, warning and stop. Because it enables maximum use of the line, the automatic block has now become the rule for lines which have dense traffic moving at high speeds.

The possibility of another train entering a block section which is already occupied cannot be totally excluded, notably in the case of accidents. There are two very different ideas in use here: 'absolute blocking', whereby such occurrences are very unusual and action is taken according to the particular case; and 'permissive blocking', whereby the train is stopped at the entrance to an occupied section and then allowed to proceed slowly, with the driver watching the line ahead until normal conditions are re-established. In practice the railway rules come somewhere between these two extremes. An intermediate solution is that of the automatic block for long sections of line, used on lines which have average traffic.

Face-to-face signalling on single tracks must satisfy both the requirements of braking distance and of spacing of trains. The simplest system, used on lines which have very little traffic, is to keep strictly to the timetable. A more elaborate system is the 'train offering', effected by the exchange of messages between stations, where a single track is doubled. It may be reinforced by interlocking this with closing the entry signals on single-track sections. This is called the 'interlocking single-track block'. A device widely used on such a block is the 'electric single-line token'. A train may not travel on a single-track section unless it carries the token for that particular stretch of line. This cannot be taken out of its storage container and given to a driver if another token for the same section has already been taken from one or other of the containers situated at the ends of the section and electrically interlocked. An important advantage of this system is that the train must slow down at the entrance to these sections in order to exchange tokens, often while still in motion; or else it must stop. No signals are needed for this, so the system is therefore especially economical for lines which have only moderate amounts of traffic.

The most comprehensive solution is centralised traffic control, whereby every signal and points manoeuvre is carried out from one box. Such installations currently control single-track sections which have heavy traffic (up to eighty trains per day) over several hundred kilometres and combine, as we shall see in the following chapter, both safety and production control.

Supplementary signalling systems are also necessary for the control of trains on some parts of the line. Permanent speed limits depend on the layout of the line and the presence of viaducts, tunnels and bridges; these may be different for different types of train. Temporary speed limits may be signalled for trains crossing sections on which maintenance is being carried out. Finally, signals may be used to modulate train speed so as to create a rhythmic traffic flow, using techniques which will be outlined in Chapter 3.

On lines which have very light traffic, such signals might be simple mobile flags. However these have rapidly been replaced by signs whose shape and colour vary according to their functions; this means they do not have to be illuminated by day, but leads to very complicated installations in the larger signal boxes. Also luminous panels are used more and more frequently. These are similar to road signs and are either lit permanently or only when a train passes. Finally, it is possible to move away from the idea of fixed signal, by directly transmitting instructions to the driver's cab, by means of cables, beacons and track circuits; this is called cab signalling.

Trains normally have a tail signal placed on the rear of the last vehicle which allows station and signal box staff to check that the train is complete. This signal may give supplementary information about the type of train (regular, relief train, etc.). Checking that trains are complete may also be done by automatic devices which count the number of axles, interlocked with signal-setting devices.

272 *Braking*

The purpose of brakes is to make the train slow down or stop in compliance with the orders given by signals, or in case of unexpected malfunctions. The maximum acceptable deceleration is limited by the comfort of the users; it is usually 1 m/s^2. Either friction brakes, acting on the wheel or on the rail, or various types of dynamic brakes are used.

With wheel friction brakes, the retarding force is usually applied by the intermediary of iron shoes acting on the wheels, and the energy dispersed by heating the wheel rim. This is the basic railway brake, using the friction coefficient of iron on steel. This varies with speed and the force applied to the shoe; iron may be replaced by composite materials whose friction coefficient is less sensitive to speed. Friction may be exerted on other places than the wheel rims; this is the principle of disc brakes, adopted by the car industry, which dissipate heat better. Some passenger trains are equipped with combined brakes using both shoes and discs.

Braking power, like tractive effort, is subject to the constraints of adhesion, which cause the wheels to lock and slide above a certain value of retarding effort. Because adhesion depends on several parameters and the trains to be braked travel at different speeds with variable numbers of empty and loaded wagons, the determination of braking distance depends on statistically-based empirical methods.

On some railways, especially in Europe, where the block sections are short and goods equipment very diverse, the complex notion of 'braked weight' is used to arrive at a satisfactory average braking efficiency.

Continuous automatic brake systems employ the action of air on pistons controlling the movement of the brake shoes through the intermediary of reservoirs and braking gear under the vehicles. Starting from a compressor in the locomotive either high pressure (air brakes) or low pressure (vacuum brakes) is created and transmitted by a braking hose, with flexible couplings between vehicles. Should the braking hose open accidentally, the brakes automatically come on. Air brakes are the most widely used, but unfortunately they do not act immediately. The time they take to work may be reduced by electrically controlling the action of the brake on each vehicle. This is the electro-pneumatic brake, which is useful on very long trains.

The electromagnetic brake uses a magnetic field created by an electric coil to apply a brake block to the rail. It is therefore a friction braking system independent of adhesion. The fierce heat generated may cause some local changes to the structure of the rails, so its use is generally limited to emergency braking.

There are several types of dynamic brakes. The first is the electric brake, either rheostatic or regenerative, which use traction motors as generators. The current is discharged either in the resistances or through the overhead contact wire. The latter is very economical on gradients; the energy regenerated on descent can be used to supply trains as they ascend (mining lines). Its use is limited by the power of the engines and the heat dissipated by the rheostats.

The hydraulic brake, similar to a hydraulic clutch, uses either water or oil. It is still in the experimental stage.

Finally there is the electric brake using Foucault current. These are rotary brakes similar to those on heavy lorries and some linear brake blocks. Heat is dissipated to the rails themselves, which poses the problem of their overheating.

273 *Safety at high speed*

The recent and continued increase in the speed of passenger trains has posed new problems in the field of safety. Different solutions have been developed for existing lines, which in the main have already been equipped with the most sophisticated signalling, and for new lines. Experience acquired during the last 20 years fixes the division between these categories at around 200 km/h.

On existing lines, usually with heavy traffic and already equipped with automatic blocks on short sections, safety at high speeds is not simply a matter of modifying this spacing for a few high speed trains. Rather, the solution is to use an extra section, called 'the advance warning' section, when the normal braking distance can no longer be observed. Initial braking at the start of the advance warning section allows the driver of a high-speed train to approach the section at the maximum speed of normal trains. In practice a combination of green and yellow lights are used, which complicates the circuits but not the control panel itself. In addition, continuous automatic brakes are usually complemented on such trains by another system for use as emergency brakes, such as disc or electromagnetic brakes.

On new lines the problem is simpler for they are designed, as is the corresponding equipment, for high-speed use. Most often only one type of train runs on them, while stations are few and of simple structure. The best signalling–braking combination can therefore be sought, within limits set by a small number of technical and economic parameters. The block principle is retained, but the length of section is increased and may be varied according to the longitudinal profile of the line. Permanent cab signalling is most frequently used. Shoe brakes do not allow speeds exceeding 200 km/h, disc brakes function up to 260 km/h but deceleration is noticeably reduced, while brakes using Foucault currents pose problems of heating. So the solution generally adopted is a combination of traditional friction brakes (discs or shoes) and electric brakes for traction wheels.

274 *Driving trains*

'Every member of staff, whatever his rank, owes complete obedience to the signals'. This quasi-military command, which figures in all railway regulations, is addressed primarily to train drivers, who carry the responsibility for several hundred human lives. The driver of a modern locomotive has a work place which provides sufficient comfort over periods of several hours, in the course of which all his attention must always be directed to observing signals, which come and go ever more rapidly as speeds increase. However human failure cannot be totally eliminated and since the operational unit of real time has been diminished to just a few seconds on some lines, various aids to the driver have become indispensable. Their purpose is to keep the driver alert, either by giving him a short manual movement to perform at intervals (a device called 'the dead man's handle') or by making him carry out precise manoeuvres when passing signals. In both cases, if the action is not performed an emergency brake is applied. Relay between the signal and the locomotive is effected by electrical contact or induction devices; it is normally restricted to warning signals. Also used are detonators which explode when a locomotive crosses a stop signal. Cab signalling makes it possible for conditions up line to be permanently displayed.

Ultimately it is possible to envisage automatic driving, requiring the running of trains to be programmed. However, because some of the parameters are variable (load and speed of train, type of locomotive, etc.), it would be necessary to have multiple programmes for any one line. There are also numerous temporary constraints, such as sections under maintenance. In practice, therefore, programmed running can only be used for trains of similar composition and speed, such as metropolitan subways, suburban trains, and block trains running on specialised lines. However, such automatic driving does not dispense with the presence of a person in the cab, who remains necessary in case of malfunctions.

275 *Staff safety*

This is a permanent concern of the railways, and the results obtained in reducing accidents at work bear witness to this. All the traditional techniques of safety training and accident prevention are used, especially in the form of periodic campaigns dealing with one particular aspect of safety. The principal areas specific to the railway are:

(*a*) maintenance sites on the track, especially on multi-track lines;
(*b*) high-voltage electric conductors;
(*c*) wagon shunting, especially in marshalling yards.

Preventive measures have to do with the construction of equipment, protective clothing for staff, and lighting and soundproofing in maintenance depots.

276 *Safety of third parties*

We shall confine ourselves to the two most important aspects. The first is enclosing the track to prevent people or animals crossing them inadvertently. The present tendency is, however, not to enclose track in the country any longer, except on high-speed lines where, just like motorways, it is necessary to prohibit access altogether. When any railway, except a tramway, runs through the centre of a town, as in North America, or through areas open to the public such as ports, the rule is to have someone precede the train on foot.

Level crossings have been the most dangerous points since the development of road traffic. The principle is that rail has priority over road, which requires road signals and some sort of protection from the railway in the form of barriers. Nowadays the most frequently used level crossings are being replaced by subways or overpasses, at times a very costly investment. On lines with average traffic, there are automatic barriers controlled by cybernetic minisystems.

277 *Development of safety*

It is apparent that safety is a field in which cybernetics has made possible, through guiding and track circuits, decisive improvements, leading to the possibility of entirely automatic traffic. The development of automation is also a powerful factor in improving output of lines and the smooth flow of traffic. The most notable developments in the field of safety have come in response to the continuing increase in speed and thus the diminishing of the operational unit of real time, which means that the driver must be kept permanently informed of what is happening down the line.

But safety is costly, especially at junctions which require special systems of stop signals. This is the case at small stations on main lines, and on branch lines where a regular passenger service must be maintained. On the other hand, the equipment needed on a specialised line, even high-speed lines where problems are limited to the spacing of identical trains travelling at the same speed, call for only a small outlay per unit of production.

28 Areas of research

After this brief survey of the present state of railway technology, I hope I have made apparent the fruitfulness of the fundamental principles of guiding and adhesion. The railway system has discovered how to get the best from them by applying all the tools developed over time by the art and science of engineering, most recently traffic control and automation. This chapter has also shown to how great an extent the technological evolution of the last 20 years has been dominated by increasing speeds, from which the spin-offs have been particularly rich and sometimes quite unexpected. There are now trains running at up to 300 km/h. Is further research necessary to increase this figure, which is surely not a technological barrier? Given the foreseeable development of the market, of the cost of energy and of possible alternatives, the answer for the moment is no, as will be shown in chapter 4.

A single field, though a particularly important one, depends on fundamental research, properly so-called. This is the behaviour of the track–vehicle couple, the key to stability, comfort and increased speed. Research here involves recourse to complex mathematical theories and to simulation techniques, as well as to detailed and costly experiments. All other areas depend on applied research. They open vast perspectives, but it is essential in the present difficult economic circumstances to channel this research towards fundamental objectives (satisfying the market, reinforcing safety, reducing costs, responding to ecological problems) and to resist the temptation of perfection and sophistication. There is still enormous progress to be made in the matter of standardisation. In an age of macrosystems this can come only through compromise and in this respect the railway still lags far behind the aeronautical, car and computer industries.

Is it possible to define an optimum framework for technological research into railways? There is no doubt that if Europe and Japan have constantly been the pioneers and remain so, this is because their main railways long ago developed specialised research services with high quality staff and installations. In addition, they often call on universities and industry to help in their research. This has been profitable in all areas and it is desirable that it continue – though at lower cost. This is why the most important macrosystems of Europe and North America have developed communal research structures. At the same time, progress in standardisation will decrease the need for research bodies and testing facilities on every railway. Substantial economies in the management of railway research therefore seem possible without decreasing its efficacy as long as the desire for co-operation can overcome some technical chauvinism, which though understandable is now out of date.

3

Production

We now come to the functional subsystem which makes technological resources available for the formation and running of trains, with the purpose of best satisfying the various requirements of demand. This, as I have said, is something specific to the railway, the only mode of transport to combine guiding and the convoy. It is sometimes called the running of the service or 'operation'. After defining the objectives, I shall examine the dispatch of passengers and goods, then the use of installations, rolling stock and staff before concluding with the structure of costs and operational management.

The principal units of railway production are: the train–kilometre (trk); the gross tonne-kilometre hauled (tkh); the passenger–kilometre (pk) and the net tonne-kilometre (tk); seat-kilometre provided (skp) and tonne-kilometre provided (tkp); and kilometre traffic unit (ktu).

31 **Objectives**

These are simple to formulate: maximise the efficiency of fixed installations, of manpower, and of the system's rolling stock, as a function of demand. There are several difficulties though. First, transport cannot be stored. Second, passenger or goods demand is as a rule individual and often uncertain, while the supply, i.e. the train, is collective or grouped and usually scheduled. Except on specialised railways, the line and part of the fixed installations are used to respond to two totally different markets, passengers and goods. Finally, for the user, the execution of a transport operation very often requires the successive use of several modes of transport. Interfaces therefore play a very important role.

The various forms of supply will be examined from a technical point of view only in this chapter. Choice among them, and their different modes, are the province of sales and management, which are the subject of later chapters.

32 Dispatch of passengers

The railway provides its users with a regular service of trains at published times, as previously did stage coaches.

321 *Train structure*

There are two kinds of train structure. The first is the block train of rigid composition. This is the kind of train used for regularly timed services because it eliminates the need for shunting at each end of the journey. It consists either of a rail motor set or of units hauled by locomotives that can be driven from both ends, and in theory allows a certain amount of standardisation among tractive units.

The other kind is the split train, whose composition varies over the journey, which permits the dispatch of groups of through coaches by several successive trains after shunting operations at connecting stations. Some small-size railways (NS) provide an excellent linear-grid service, linked by means of split trains each made up of rail motor sets that can be shunted very quickly in specially designed connecting stations.

Technology has made it possible to provide trains whose capacity varies between 100 passengers in single railcars and 2000 passengers in double decker trains. Maximum length is determined by the installations at stations; it is about 200–300 m for metropolitan subways and may reach 500 m for intercity trains. Capacity is dependent on the level of comfort. A train 400 m long can carry up to 1000 seated passengers, i.e. 2.5 per linear metre. This figure may go down to 1 on high-comfort trains.

The composition of a train is adapted to fluctuations in demand in several ways. Trains which are used by daily commuters (metropolitan and suburban services) usually have a fixed composition and rely on frequency and a large number of standing passengers. For intercity connections which use block trains, it is necessary to reserve a seat beforehand, as for aeroplanes or long-distance coaches. On the other hand the composition of traditional intercity trains very often allows one or two extra carriages to be added, amounting to 10–15 % extra capacity, without changing the timetable. Some railways systematically modify the composition of their trains according to the days of the week, but the usual solution is to run relief trains. On a large number of railways, there are two timetabling services, one corresponding to a basic service, frequently called the winter service, while the other is significantly heavier, corresponding to the normal holiday period, and called the 'summer service'. This lasts four months in Europe. These periodic supplementary trains are sometimes faster than the normal trains and also run over short periods out of season, which ensures that full use is made of stock.

322 *Train speed*

As we saw earlier, the potential speed of passenger trains may be 200 km/h on some sections of existing lines and 300 km/h on specialised lines (the Shinkansen and Paris–South East).

The commercial speed depends on the number and length of stops. The 'speed efficiency' thus varies to a considerable extent. For the Paris–Brussels TEE (312 km non-stop) and Hikari trains of the Shinkansen between Tokyo and Hakata (1177 km with 6 stops), the efficiency is 0.8, and this seems to be the maximum. On some long-distance split trains however, the efficiency may fall to 0.5.

323 *Interfaces*

The passenger station is the interface between the railway and its users. Its design is a response to technical and commercial needs.

Technically, there are two types of station structure. A station for stopping trains has one platform for each track. This is the standard form for metropolitan and suburban lines and is also used on some specialised high-speed lines.

The second type of station structure provides for modifications in train composition and for service connections. It includes one or several groups of tracks, either through lines or dead-end lines, as at terminal stations. Some large stations have more than 30 tracks (Leipzig, New York Grand Central). Preparation, minor maintenance and storing of trains, together with sidings for reserve stock sometimes make it necessary to have enormous groups of tracks near main line stations.

In the commercial area, every station has the necessary offices for information, buying tickets, reserving seats, registering and reclaiming luggage, and there are also waiting rooms and restaurants. The development of the air terminal has made it necessary to improve the level of comfort and the amenities provided at railway stations.

Many stations are situated in the centre of town and very often were built at the same time as the railways themselves. This is a very significant advantage for the users, though some cities have unfortunately failed to recognise this and have removed their stations to the outskirts of the city, leaving the town centre totally oriented towards the motor car. This tendency, which does the community a disservice, has since been halted. In contrast to airports, which as enclosed areas outside the city may take on the ambience of a prison, the railway station can provide a forum for social communications and a site for commercial activities such as shops, restaurants and other services, and even for exhibitions and theatrical performances[1]. The development of such functions of the station is particularly significant for the interface between the railway and other modes of transport, in the case of urban and suburban transport at the level of metropolitan macrosystems (described in chapter 6). Crossing these interfaces greatly reduces the average speed from the beginning to the end of the journey and increases passenger discomfort. Because it is often impossible even on metropolitan subways to provide connections on the same level, vertical and horizontal mechanical connecting devices, such as lifts and escalators, are necessary. Also the creation of car parks at suburban stations is one of the most effective ways of fighting road traffic congestion in the city itself. The present tendency towards conurbations has led to the creation of satellite towns. This often justifies main line trains stopping at stations serving these towns and helps reduce congestion at central stations.

The interface between the railway and the aeroplane has been improved considerably in recent years with the construction of railway lines from town to airport. Wherever possible, an airport should be served by long-distance trains, for example Berlin – Schönefeld, Frankfurt, Zürich. On the other hand, the interface between the railway and shipping has practically disappeared for intercontinental journeys because these are now made by air. However, there are still a few continental connections serving low cost and seasonal traffic, for example, between the United Kingdom and Continental Europe, which are served by maritime stations.

33 Dispatch of goods

The railways offers two forms of goods dispatch: the block train and the single wagon.

331 *Dispatch by block trains*

There are two types of block train, and their common characteristic is that they are not modified in any way during their journeys.

The commercial block train, sometimes called the complete train load, is a homogeneous convoy transporting merchandise of one kind and travelling between consignor and consignee. It has been used since the beginning of the railway and remains the sole form of dispatch on specialised mining lines. When it runs between private sidings, as is generally the case, the complete train load gives door-to-door service. These trains usually have very high tonnage; up to 5000 tonnes gross in Europe and more than 20 000 tonnes gross on some mining lines. They are specially designed for high capacity, up to 5–6 tonnes net per metre for the most recent iron ore wagons and they may be longer than 2000 m. The capacity of such 'cargo trains' is considerable: a single train of 3300 tonnes net running 300 days per year dispatches one million tonnes of goods annually. Ten daily trains of 10 000 tonnes net brings this figure up to 30 million tonnes annually.

The technical block train, sometimes called the through train, is composed of separate wagons or groups of wagons from different consignors. It is made up at a dispatch station or a marshalling yard and run without any modification to the destination station or to another marshalling yard. Examples of this are trains carrying perishable goods, running between a station serving a production centre and a station serving a market, and trains linking two terminals for large containers. Empty wagons are also moved in such trains. Through trains are lighter (1000–2000 tonnes generally) but less homogeneous than block trains; their formation and length are often limited by the length of sidings.

Dispatch by block trains does not create any special problems. They only require technical stops (to change locomotives or staff) and the commercial speed is high, especially if they have the advantage of priority over other trains. It may reach 50 km/h for ore trains and 100 km/h for trains carrying perishable goods or containers, and this over distances sometimes greater than 1000 km. The speed efficiency is satisfactory. The block train thus provides a particularly simple and effective form of dispatch.

332 *Dispatch by single wagon*

Each wagon or group of wagons is incorporated into a series of trains according to a program called 'routing'. This method usually involves three stages, separated by shunting operations carried out at specialised yards.

(*i*) The wagon is picked up by a local train serving the collection point and is taken to a marshalling yard (the workings of the marshalling yard will be examined later).

(*ii*) The wagon is transported by one or several through trains which link the marshalling yards at the starting point and destination.

(*iii*) The wagon is distributed by a local train made up in the marshalling yard serving the destination point.

The first and last stages constitute the terminal operations. Such operations which derive directly from the principle of the convoy are characteristics of the railway; they are similar to the distribution of mail by means of mailbags.

The development since 1945 of operations research and simulation techniques has permitted, after a century of empirical approaches based on an ever-deepening acquaintance with the facts of the problem, the first scientific approach to optimising routing, that is to say, transport on grid networks, the only network structure which can offer a choice of routes. A great number of studies have been carried out, unfortunately most of them unpublished, and quite good results achieved on a few railways. In particular, these studies have the merit of highlighting the fundamental problem: how to quantify the efficiency function. The formulation of this function is simple: how to maximise the quality of service and at the same time minimise cost. However, this function is practically impossible to describe in quantitative terms. Quality of service cannot be measured in the same way by all who use it. For some people the speed of dispatch will be of prime importance, for others its reliability. Then again, cost, as we shall see later, is itself a combination of individual costs: the cost of terminal operations, of shunting; of running through trains between shunting yards, of engines and rolling stock kept idle; etc. It is not possible to isolate a single variable to be minimised. On an existing network – the general case – it is impossible to abstract from the ensemble of shunting operating methods, geographical position and maximum capacity. In any event because of the necessary complexity of organisation involved in the transportation of single wagons on the larger railways, and because all those concerned with its execution must be kept informed, it creates too many problems to modify the routing too often. The tactical unit of real time is the period of the timetable. On the majority of railways it is dependent on the passenger train service, which changes once or twice a year.

Terminal operations are carried out by local trains. These have to perform a large amount of shunting, and thus they travel at very low overall speeds. Their timetables must take into account the working hours of those who use them, which usually means that they run between 6 p.m. and 6 a.m. Often the same train carries out both collection and distribution. Terminal operations for clients with small loads only, and dispersed in large industrial areas or along the length of main traffic routes, is a very slow and costly business.

There are two ways of moving goods in single wagons. The first is slow routing, where the main objective is to minimise the actual distance covered by the trains. This is achieved by linking the shunting operations with through trains of high minimum tonnage, which therefore travel infrequently. These wagons will then normally be shunted into intermediary sidings, and the stop-over will last for several hours.

With fast routing however, the main objectives are to minimise the length of time of transport and to secure quick delivery by increasing the number and the speed of through trains, decreasing their tonnage, and creating fast connections with marshalling yards. Terminal operations stages are sometimes performed by local passenger trains.

Many railways use both methods conjointly, although often with a clear preference for slow routing. This means that some have to use two separate marshalling yard grids. In addition, in a number of cases the system has to take into account 'special dispatches' which have a fixed delivery time.

The problem of random fluctuations in traffic may be solved in various ways. On some railways, particularly in the US, the through trains used between marshalling yards for slow routing are optional and run only according to demand. Such a proposition is commercially viable only when the average distance goods are transported is very high, about 1000 km. But in the great majority of cases, there is a regular framework of trains which run between marshalling yards with a minimum of one or two daily connections. Thus optional trains will be used only to absorb extra traffic at peak times, and these run if possible before the regular trains so as to clear the marshalling yards. Fast routing requires a regular service of trains running according to a strict timetable, similar to that for passenger trains. In extreme cases, a supplementary train must be used to dispatch a single wagon to ensure guaranteed delivery dates.

Specialists in routing work out a system of improvements for each timetable period by taking into account the changes in the flow of traffic and the available technical equipment. The two factors taken together mean that slow routing does not at present attain a commercial speed, from starting point to destination, of 10 km/h, even on railways where through trains used between marshalling yards have an average speed of about 80 km/h. The speed efficiency is therefore about 0.1. Obviously, this is due to the length of stop-overs in the marshalling yards and the time taken by terminal operations. Thus slow routing practically excludes the possibility of delivery the following day, whatever the distance. With fast routing the efficiency often reaches 0.2 or 0.3. Through trains have an average speed of 100–120 km/h, stop-overs are shorter and the commercial speed from starting point to destination may exceed 25 km/h. In all cases however, the terminal operations are the main reason for these somewhat disappointing results.

These figures are average values valid for large grid railways functioning as part of macrosystems; sometimes further delays may be caused at borders. The figures are perceptibly higher on railways which have a linear or branched structure.

333 *Development of dispatch methods*

The intrinsic slowness of goods transportation, its poor quality as compared to road transportation, the development of demand, and finally the possibilities offered by intermodal techniques – which we shall examine in the following chapter – have led a large number of railways to think again about the whole problem of goods dispatch. Not enough work has yet been done on this and to a great extent each network is a special case. However a few general solutions are beginning to emerge.

With block trains, the objective aimed for up to now, and which is still valid for bulk transport, has been to maximise the tonnage of trains. Originally, these trains, especially complete train loads, were not meant to have a tonnage of less than several thousand tonnes. But many networks recognise the existence of regular demand of the order of several hundred tonnes. Routing meets this demand poorly especially if several shunts have to be made in transit, for the original group of wagons rarely reaches its destination at the same time. A number of these railways have, after detailed economic studies, reached the conclusion that the mini block train of around 500 tonnes is an economically acceptable solution and also commercially attractive. Indeed, it considerably improves the efficiency of equipment and in some cases makes for substantial reductions in the number of crew. Finally, it requires no capital investment.

Another recent development is the semi through train, a combination of block train and fast routing. It involves dispatching through trains composed of two or three groups of wagons assembled in simple formation sidings without individual shunting, following the technique used in passenger stations for transferring groups of through carriages to connecting trains. This formula can only be used after a detailed analysis of traffic flow, but its advantages are equivalent to those of the mini block train. It does not systematically eliminate the initial and final shunting.

334 *Interfaces*

Local goods yards, which were very often linked with local passenger stations and therefore located on urban sites, are now disappearing. Their facilities are generally not large enough for complete train loads, but above all the development of distribution methods has considerably reduced the need for them. Such yards, however small, are also costly to build because of their safety equipment and costly to run because of the slowness of shunting. However, they do often lie on very valuable land. The present tendency is to maintain local yards only, serving reasonably large towns by wagon load or groups of separate wagons, but without developing their capacity.

Instead the orientation is towards the creation of centralised dispatch or reception yards which are specialised for particular traffic and capable of forming or receiving block trains: agricultural markets, industrial zones, inflammable goods and materials depots, etc. Some of these yards need small shunting facilities. They may be built outside city perimeters, which is an advantage for the environment.

However the essential interface is still the private siding, initially conceived for bulk traffic to mines, quarries and factories. When competition from the roads appeared it led the majority of railways to practise an active policy of creating private sidings through financial incentives. This policy has borne fruit because the traffic dispatched between private sidings is more than 80 % of the total traffic on some networks. However the service to private sidings which have little traffic poses the same problems as those of small yards, problems which are aggravated when the sidings are on open lines and are accessible from one direction only.

The development of the two principal intermodal techniques, the large container and road trailer (see chapter 4) has led to the construction near large urban centres and in ports of specialised termini for loading, unloading and storage of these transport units. Some harbour termini deal with up to 2000 large containers (20 foot units) per day. These installations, which require vertical handling methods and take up a lot of room, cost a great deal of money and their number is limited. Very often they belong to organisations outside the railway. Termini for road trailers are very simple, consisting merely of access ramps for wagons and to storage areas.

34 Utilisation of fixed installations

Because of guiding the position of a train at any instant is definable if its speed and the geography of the route are known. The basis of the utilisation of the physical railway network is therefore a space-time diagram called a 'train diagram'. Its best use depends first of all on the timetable, which I shall examine first before looking at the problems of capacity which arise in various ways depending on the types of line and junctions.

341 *Timetables*

The problem has two levels for any given section of line:

(*i*) determining for each category of train the 'standard running';

(*ii*) optimising the combination of the different standard runnings to satisfy a demand which is both quantitative (number of trains of each category) and qualitative (positioning of certain trains according to the needs of that time of day) so as to obtain the train diagram.

The standard running of a train involves the parameters concerning the geometry of the line (slopes, curves, limited speed zones etc.) and the dynamics of the train (resistance to motion, the tractive effort curve and speed of the locomotive normally attained under haulage, acceleration and braking performance). Also chosen is the best regime of use of the locomotive, a compromise between tonnage and speed, and a margin of security is added to take into account the possibility of limited overloading, or track maintenance and traffic accidents. The number and length of stops are fixed according to commercial needs (passenger movement, connections, etc.) and technical considerations (changing locomotives, changes in train composition, formalities at borders between countries). The calculation of the standard running, which makes use of mathematical integration, is often carried out by electronic calculators. Equivalence tables make it possible to adjust the tonnage supplied according to the type of locomotive available.

The train diagram is established for a period of 24 hours. It involves an additional parameter, the minimal delay between two successive trains, itself dependent on the spacing between signals on the line, and stopping times for slow trains to let faster trains overtake and to let trains cross on single-track lines. On specialised lines where only one category of train runs (metropolitan subways, regional and intercity passenger lines such as the Shinkansen, mineral lines) the train diagram is reduced to segments of parallel straight lines. It can be broken up into repetitive groups of successive trains making part or all of a journey with a number of different stops; the rhythm of the group may vary between 15 minutes and one hour, and, as a general rule, no train inside the group can overtake another. But on lines which have mixed passenger and goods traffic, the establishment of the train diagram involves the definition of a hierarchy among the trains. The majority of railways give priority to intercity passenger trains and to certain block trains, whose timetables are fixed by commercial considerations. Working out the train diagram then becomes very complex and can only be the result of compromise. In practice most railways try to make local improvements each time the service is changed.

Train diagrams often include, as well as the times of regular passenger trains, the times for temporary relief trains and for optional goods trains, whose unpredictable running is thus made easier. Extracts from train diagrams are distributed to the train crew and to stations, signal boxes and traffic control centres. Finally the train diagrams are used to compile the timetables for passenger trains and for goods trains on schedules with guaranteed delivery to clients.

The duration of a train diagram defines the tactical unit of real time of railway production, on which is based the use of rolling stock together with part of the personnel. The establishment of the train diagram is a long and very detailed job because of the close concern with safety; it is therefore costly. On a large railway, especially if it is part of a macrosystem which implies prior international agreements, it normally takes several months from the time work begins to the moment the documents are handed out. So it is sensible to increase the length of time they are used. Most railways adopt a yearly cycle, which allows extra trains to be run at seasonal peaks, and the publication of two timetables per year for its clients.

On the other hand, the increase in speeds has led to changes in the operational unit of real time, at least on some lines. The old unit was normally the minute, but in one minute a train travelling at 200 km/h covers 3.33 km and crosses several block sections. It has therefore become necessary to operate with smaller units. Also, greater use is being made of computers which make it possible, by using simulations, to evaluate the results of trains travelling at different speeds and at different times, to locate any incompatibilities in traffic and to publish the finished timetables more quickly and accurately.

342 *Double-track lines*

It is apparent from what has already been said that the capacity of a line is a relative notion. Maximum capacity can be defined only for a line with a single standard running or train group. It then depends solely on the minimum interval between two trains or groups. For most other lines this idea only makes sense if it is linked with a pre-determined program for trains which have priority; one then tries to maximise the number of runnings left available for other trains in the intervals. There are operations research techniques which can help in obtaining solutions, given that there is only one standard running for non-priority trains; these techniques, which involve the use of computers, are also used to determine the consequences for the capacity of the line of running an additional priority train. The determining parameter is the speed differential between the standard runnings. A train travelling at 200 km/h on a line several hundred kilometres long greatly reduces the maximum number of trains running at 80 km/h. This shows the desirability on lines with heavy traffic of increasing the speed of slow trains and the cost to capacity of high speed trains on non-specialised lines. Also, on systems which give priority to goods traffic, passenger trains can reach speeds which merely equal or are only a little faster than those of the goods trains on lines with heavy traffic.

Finally, the real capacity of a line must take into account track maintenance particularly on electrified lines. On lines which have little or no night traffic, such as metropolitan subways, regional and intercity passenger lines, such repair work can be carried out at night, and capacity is not reduced. But for normal lines, some railways set aside certain periods for the maintenance of overhead wires. The length of time varies between one and two hours, which reduces actual capacity.

The capacity of a double track line can be increased by successive phases, working on making use of the characteristics of the trains, their distribution throughout the day, and the fixed facilities. The normal sequence is like this:

(*i*) increasing the tonnage of the trains and the acceleration/braking performance (more powerful locomotives or multiple units);
(*ii*) increasing the speed of slow trains;
(*iii*) grouping trains of the same speed in batches with a minimum interval between them during certain periods of the day (night passenger trains, etc.);
(*iv*) ending local passenger services;
(*v*) installing short-section automatic block signalling;
(*vi*) providing sidings areas with direct entry for slow trains, at regular intervals;
(*vii*) creating signalled track sections for two-way running, or three tracks, with the third track being used for two-way running, distributed according to the rhythm of the train diagram;
(*viii*) finally, by doubling the line, with the same characteristics, or by constructing a new line specialised for fast traffic, in the form of a corridor (see below).

It should be noted that two-way signalling, making it possible to use each track in both directions on a double-track line, is principally a means of easing the traffic flow when there is maintenance work or an accident on one track. When a line is near saturation point, allowing two-way traffic will not increase its capacity. On the other hand, it is useful when, at certain times of the day, the density of traffic is very different on the two tracks.

The experience of railways with different train priority systems has yielded similar results: the maximum number of trains on the two tracks over a 24 hour period is 250–300, with the possibility of peaks for a limited period. On a metropolitan subway line with an interval of 90 seconds, 40 trains can run every hour in each direction, or 720 trains in the period 6 a.m.–12 midnight. On a main line with a three minute interval, groups of 20 passenger trains can run every hour.

343 *Single-track lines*

With single track an extra parameter comes into play: the travelling time between overtaking points, usually the longest time section on the whole length of the line. As a result the capacity of a single-track line is necessarily lower than that of each single track on a double-track line. Further, it is much more susceptible to delays from maintenance and accidents because all changes at intersections involve chain reactions. Methods for gradually increasing the capacity of a single-track line are, in sequence,

(*i*) increasing tonnage and train performance;
(*ii*) increasing the speed of slow trains;
(*iii*) creating groups of trains travelling in the same direction at certain periods of the day;
(*iv*) providing overtaking points which equalise travelling time, taking into account the geometric parameters of the line;
(*v*) providing extra overtaking points;
(*vi*) centralising control of signalling and traffic;
(*vii*) doubling track on particularly difficult sections and in long tunnels.

Experience shows that it is very difficult to exceed 80 trains per day, travelling in both directions, on a long single-track line with both passenger and goods traffic. On mining lines, taking into account the possible tonnage of complete train loads, single track will remain sufficient for a long time to come. On the other hand, it is very difficult to provide a fast passenger service of sufficient quality on single track.

344 *Corridors*

The optimum use of this topological structure makes it possible in many cases to improve the capacity of a railway route over several hundred, even several thousand, kilometres through a mixture of single or double-track sections. The guiding principle is to reserve the best route in the corridor for express passenger and goods trains, with normal goods transport using other combinations. Each corridor is an individual case, with extreme examples consisting of lines running in parallel for several kilometres, (the Rhine or Rhône valleys for example), or separated by several hundred kilometres (the Trans-Canadian or Trans-Siberian on some sections). The idea of the corridor has only proved of value following the construction of lines conceived for high speeds (Europe and Japan), or the remodelling of passenger services (US, Canada). It may well find a role at the level of macro-systems.

345 *Passenger stations*

The use of passenger stations for purposes other than simply as stopping places presents the problem of optimising the set of tracks. This is an important problem, for the corresponding financial investments are very high (occupation of land in urban areas, points and crossings, safety devices), and they are difficult to modify, particularly on electrified lines. The objective is to minimise the number of platform tracks. The solutions vary according to the type of service (rhythmic or variable period) and the configuration of tracks (through station or terminus).

For a terminus station on a line with specialised and rhythmical traffic, the fundamental parameter is the stabling time between arrival and departure of the same train. On short-distance lines (metropolitan and suburban) the train can be driven in either direction and no technical or cleaning operations are necessary. The stabling time can thus be limited to a few minutes. If its length is less than the traffic interval, a single track is theoretically sufficient for each line. If not, two tracks are necessary. The basic equipment is therefore very economical. However, normally another track must be added to deal with fluctuations and accidents.

On intercity lines with a rhythmical service, the stabling time must be long enough to allow the train to be cleaned, food and water supply to be replenished, reserved seats to be identified, etc. This takes a minimum of 30 minutes, which is usually longer than the traffic interval. In such a case, four tracks are normally necessary. The Shinkansen terminus station in Tokyo has five tracks and three platforms, which receive and send out 262 trains per day between 6 a.m. and 12 midnight. The station is just 56 m wide though it deals with traffic which reaches 162 000 passengers daily. This illustrates the enormous superiority of the railway as compared to other forms of transport as far as the efficiency per unit of land is concerned.

In these two cases, the cul-de-sac design is usually adopted. However terminus stations on metropolitan subways usually also give access to the repair depot, which it is more economical to locate underground. In all other cases, the present tendency is not to construct any more cul-de-sac stations, because they take up too much room and slow down the formation of trains. The number of platform tracks depends on the number of destinations served by trains, on the rhythm of trains at peak times, and on the length of stop judged necessary before the train departs, particularly for long-distance trains whose capacity may be more than 1000 passengers (stops of about 15 minutes). There are operational research procedures and computer simulations which can give the best use of a configuration, taking into account the parameters of the service to be supplied.

When a large city is crossed by a long-distance railway route (Berlin, Brussels, Madrid, Warsaw), the tendency is to reserve the station in the centre just for the passing of through trains, with one platform for each direction and very short stops. Technical operations are carried out at two outlying stations. Trains from each direction can thus serve the whole of the city.

The managing tool for a station is the track occupation diagram, a large-scale version of the part of the train diagram dealing with lines served by the station. A controlling body usually linked to the station's central signal box ensures that this is respected and is adapted to cope with fluctuations in demand and technical hitches. Automatic identification devices on the trains make possible in the most recent installations, continuous real time control of traffic in the approaches to the station.

346 *Marshalling yards*

The marshalling yard, a consequence of the convoy, is a largely autonomous subsystem of costly construction and operation. A large modern yard may occupy an area of 500 ha, have 250 km of track, and require more than 1500 staff to perform more than 5000 wagon shunts per day. The optimum use of marshalling yards has therefore become an important field of application for railway operations research.

A marshalling yard is made up of groups of tracks with clearly defined functions: reception, splitting up, formation, and waiting for departure (figure 7). The groups of track usually have a double entrance, allowing the use of the two ends for different functions (for example, splitting and forming trains). Shunting is carried out by gravity. The number and size of yards depend above all on the main function of the shunting, whether for transit or local service. Simple transit does not require the formation of trains for departure, while simple local service (to industrial areas, ports) is limited to this function. In addition, some yards are used to bring empty wagons together, which means specialised tracks are necessary. In practice, the reception and departure areas, which are not used in fast routing, rarely consist of more than about ten tracks, and tracks used for splitting and forming trains (often combined) number about 50. When flow and volume of traffic justify it, two adjacent shunting yards may be constructed, each dealing with traffic in one direction only, and linked by connecting tracks.

The design of a marshalling yard must take many parameters into account: the number of tracks in each group, the speed and independence of inter-group traffic, whether there is a locomotive depot and a

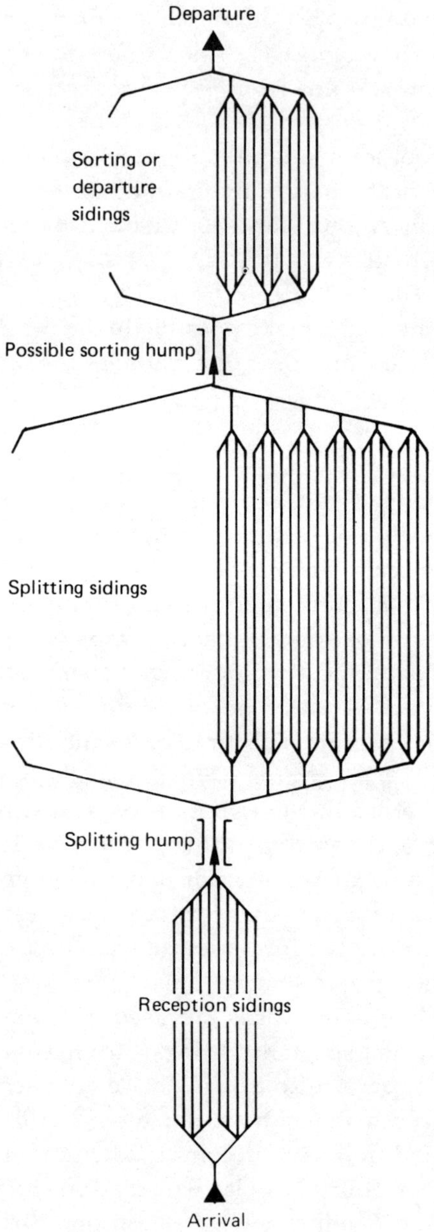

Fig. 7. Schema of a marshalling yard

workshop for quick repairs to equipment, etc. Further, marshalling yards are located away from urban areas because of the noise.

The running of a marshalling yard seeks to maximise the use of installations for splitting up trains because this is the most costly. Elementary operations, simple and repetitive, lend themselves very well to automation. They follow each other schematically in this sequence:

(*i*) prior information on the composition and destination of wagons making up a train is received (often by teleprinter message) and a plan for splitting the train prepared in accordance with the plan of track allocation (which may vary throughout the day);

(*ii*) a control check of train composition is carried out on arrival;

(*iii*) the train is pushed to the hump (by remote control of the shunting locomotive);

(*iv*) the train is split according to the destination of the wagons (automatically if coupling is automatic);

(*v*) each cut of wagons is sent to its proper track by pre-assigned automatic points;

(*vi*) the speed of each cut is regulated according to track occupation, so as to make it stop without shocks that might damage the load and without intervals between cuts. This 'precision firing' is effected by means of cybernetic minisystems, such as brakes on the tracks and accompanying devices on the wagons.

It has become possible to split up to 180 wagons per hour with a very small staff (six people) and using only two shunting locomotives. During certain periods, trains are often formed at the other end of the splitting-up tracks. Mathematical combinatorial analysis provides methods of minimising the number of shunts for a group of departing trains. The composition of these departure trains is transmitted down the line, most often by teleprinters. The importance of information and automation in running marshalling yards can be appreciated when we consider that the operational unit of real time, determined by the rate of splitting, is of the order of 20–30 seconds.

The management of marshalling yards is aided by a control unit with a diagram of track occupation in the different groups and of traffic between groups. These units are linked to control posts on the lines serving the yard in question as well as other nearby yards and locomotive depots. Techniques of operations research and computer simulation studies make it possible to construct working models of marshalling yards, according to certain criteria (waiting time of wagons, number of splitting tracks, etc.). These are used either to determine the layout of a new marshalling yard, or to improve the use of an existing yard. Given the number and interdependence of parameters, they do not claim to provide optimum solutions, but they are valuable instruments.

347 *Track maintenance*

After a long period in which it consisted of repairs analogous to those carried out on roads, maintenance of the railway has become essentially preventive. It is based on cyclical checks on all track components, carried out on sections several kilometres in length.

Periodic checking of the state of the track is an indispensable operation not only for safety but also for determining which sections will require maintenance next time round. Once manual, this checking has become largely automatic, due to the use of recording coaches containing devices measuring different parameters of track geometry (levelling, superelevation, gauge, etc.). An analysis of these recordings makes it possible to discover more about the behaviour of the track when in use. It illustrates the importance of low-frequency vehicles.

Maintenance of track includes levelling operations and the replacement of worn components. Levelling is carried out mechanically, either by packing (the efficiency of packers can be as much as 1 km per hour), or by blasting, a more sophisticated technique which judges the quantity of ballast to put under each sleeper.

Track renewal is a very important operation, which is only begun when the track is so worn that simple maintenance is no longer enough. The period of renewal depends on the amount of traffic on the line. At the moment several hundred million tonnes gross are permitted, which correspond to at least ten years traffic on the most heavily loaded lines. Track renewal requires a lot of equipment but the operation is completely mechanised and more than one kilometre can be renewed in a day. Renewal has serious consequences for the running of the service (single-track sections, speed limits), and must therefore be carefully concerted well in advance with service operation, so as to coincide with periods of slack in the commercial timetable. Some railways contract out renewal work to specialist companies with their own mechanical equipment.

35 Use of rolling stock

351 *Principles*

Rolling stock represents an extremely important investment and it is therefore essential to optimise its use. The analysis of passenger and goods dispatch distinguishes two techniques: regular dispatch, which implies a programmed use of rolling stock, and optional dispatch, which implies irregular use. The great majority of railways apply both techniques simultaneously in varying proportions and partly with the same stock. Also, a train running regularly between marshalling yards, for example, combines the programmed procedure for the locomotive and the irregular procedure for the wagons which compose it.

The organisation of rolling stock used for regular dispatch is effected in a simple way by space–time diagrams or rosters based on train compositions and timetables, together with technical constraints on stock (for example maintenance). The duration of a roster normally corresponds to that of the timetables. Computers are now being used to form trains and improve these rosters; the objective here is to maximise distances. Optional utilisation of rolling stock, which particularly concerns goods stock, poses difficult problems. I shall look at engines first, then passenger equipment and finally goods equipment.

325 *Engines*

The disappearance of steam traction, which for a long time linked locomotives to one driving crew, has meant that engines can be used at least twice as much as before. Electric traction does not require frequent stops at fuel depots, and driving crew can be changed during stops at stations. It is possible to operate diesel or electric locomotives 1000 km from their home depot, making complex circuits around grid networks. These engines can often be used equally well on several types of train, for example intercity passenger and fast freight, which increases their efficiency. Specialised lines usually use a single type of engine; this is also the case on some systems whose passenger traffic does not require high performance.

The optional service is responsible for dealing with variable sections of block trains, through trains, and peak traffic passenger trains. Advance notice of some hours is necessary to get the best use from the engines and to arrange crew. Procedures differ according to the dispatch policy and the topological structure of the networks. When an optional service is used very frequently, it may be necessary to use operations research and computers. The control centres of optional trains are usually situated in big marshalling yards, where an information network is available.

On grid railways with high traffic density, annual distances of about 300 000 km for fast passenger engines can be obtained. Specialised high speed lines have made it possible to cover up to 2000 km between 6 a.m. and 12 midnight. Some maintenance operations, such as re-profiling wheel tyres, can even be done during stops at terminal stations. While the annual distance covered by steam locomotives rarely exceeded 50–60 000 km on average, diesel and electric locomotives do more than 100 000 km and sometimes reach 200 000 km. Least efficient are rail motor sets used on local services, which only run by day. Engine stock is managed individually, usually by its home depot, which means that maintenance can be planned.

353 *Passenger stock*

The composition of regular trains, fixed for the duration of the timetable period, may vary according to demand. Additional reserve coaches or sets are stored in big centres in case of accidents. Coping with peaks of traffic necessarily lowers the efficiency of passenger stock. During super-peak periods lasting a few days, which only occur on a few Western European railways a large stock of older, less comfortable coaches may be mobilised. These are costly to maintain and involve running almost as many empty trains as commercial ones. This is a problem of business policy.

Metropolitan subway trains cover an average annual distance of 80 000 km, coaches 100 000–200 000 km, and multiple-unit electric trains up to 500 000 km (on the Shinkansen line). This confirms the importance of high speeds for stock efficiency, as the aviation industry proved when it changed from propellor aeroplanes to jets on intercontinental flights. Sleeping coaches are usually used more than day coaches; this is particularly true for stock on intercontinental trains, even if their commercial speed is quite low.

354 *Goods stock*

This problem is much more complex for several reasons:

(*a*) stock use is very often by individual unit and is irregular;

(*b*) there are many different types of wagon and their choice rests with the client;

(*c*) loading and unloading are carried out by the clients, and are often of uncertain duration.

The size of the railway, its topological structure and the constancy of traffic greatly influence the solutions. Wagons, frequently specialised ones, used to make up complete train loads call for the same techniques as passenger coaches. The complete train load returns empty immediately after unloading; though they therefore only run at 50 % capacity, turn around is very fast. In this way the mining railway of Mauritania, for example, carries an annual traffic of 12 million tonnes over 650 km with a stock of 1080 wagons of 80 tonnes net. Annual output is therefore about 180000 km per wagon, including reserve stock.

A more difficult and more general problem is that of the single wagon running on a grid railway. First of all, a wagon of a specific type must be made available to the client at a station or a designated private siding on the correct day. The efficiency of distribution of such stock is optimised by minimising the distances covered while empty, and maximising use of stock capacity. Such distribution is based on information both estimated (advance notice of demand, probable resources of local unloaded wagons) and actual (immediate review of the stock). It is necessary to have stores of empty wagons distributed among carefully chosen junctions (marshalling yards in particular). Operations research methods and computers have made it possible during the last 15 years to build models of quite complicated procedures. However these are only really effective when traffic flow is reasonably constant. Usually wagons are distributed to the region by daily trains departing from the marshalling yard, with adjustments made by trains of empty wagons running between yards.

The ratio between journeys when loaded and journeys when empty varies between 60 and 70 %; the growing specialisation of wagons tends to diminish it, but offers other advantages. The length of terminal operations is fixed by tariffs and is at the client's expense; usually it lasts a half or one day. Financial incentives (premiums or demurrage) are usually provided but their strict application is not always commercially possible.

The utilisation of goods stock is low; it rarely exceeds 100 km per day, or 36000 km per year. But these averages, the only ones to be published usually, mask great differences according to type of wagon, season (notably for agricultural traffic), and volume of traffic, which periodically leads to under-employment of part of the stock. The length of stock depreciation time prevents adaption to meet short term developments in traffic and, because transport is not storable, a surplus of equipment is always necessary. The difference between these figures and the average speed of dispatch illustrates the importance of storing wagons, be it with clients, in maintenance yards, or in sidings for stock temporarily in excess of demand.

355 *Maintenance of rolling stock*

The complete rationalisation of maintenance faces two difficulties, already mentioned in this and the preceding chapter: large ranges of identical stock are rare, and part of the stock is used optionally and (as we shall see in chapter 6) is liable on macrosystems to be used for long periods far from its own railway.

However, with the exception of goods stock on such railways or macrosystems, it is possible to follow the utilisation of each engine or vehicle by means of the data banks in operation on most big railways. Various cycles of maintenance are defined for changing main components, based on distance travelled or hours worked. Modern electric traction in particular makes very long cycles possible: several million kilometres between major checks. Some railways contract out part of the maintenance of their rolling stock to manufacturers or to industry.

36 **Personnel**

361 *General*

The railway remains, in spite of developments in automation, a major employer of manpower. At present there are more than 10 million railway workers in the world. Expenditure on staff is often considerably more than half the operating expenditure of a railway. Staffing problems are therefore at the very heart of production and management.

Since the beginning, the railway worker's job has been very responsive to the idea that it is a public service. Literature, art and the cinema have popularised numerous aspects of it. The disappearance of steam traction and the use of technology widely used in other fields have, of course, reduced its individuality, but the mystique of the railway lives on.

Railway personnel have to obey the same demands made by all transport industries: the priority of safety, continuity of service, travel, settling in. But the railway is the only transport system which is completely integrated, since it administers both its fixed installations and rolling stock. This widens the field of essential specialisations. The size of railways means that there are often workforces of more than 100000 people; the largest three, in China, USSR and India, each comprise more than a million and between them employ half the world's railway workers. This creates special problems for management. The situation is very different according to the railway and I shall simply identify general problems and solutions.

362 *Legal status*

Nearly all railways are, as we shall see in the chapter on management, nationalised industries. The status of personnel is however very diverse. In a few countries, railway workers constitute a separate branch of the civil service, with its own particular regulations. In others, a railway worker's status is that of a member of a nationalised or semi-nationalised industry. Most often a railway derives from former private companies, though now subject to rules different to those of traditional enterprises because of the constraints of being a public service. In all nationalised railways a worker has a guaranteed job. Pension rights are usually obtained more quickly for drivers. In several countries, the private railway companies were pioneers in introducing pension and insurance schemes, and a large number of railways have made their own medical services available. Private railways, which are mainly to be found in the US, follow industrial law, which excludes guaranteed employment. Some activities, such as catering, hotels, handling, cleaning of stock are not specific to the railway; many railways subcontract this work to private companies.

363 *Recruitment*

Railways usually recruit their staff on three levels: operations, supervision and management. Statutory rules sometimes hinder the recruitment of specialists in new techniques (for example computer science around 1960, or marketing). In some countries, where there is only one pension scheme, changing from the railway to industry or the public service and and vice versa is possible in the course of a career, and is usually effective. The level of salaries, especially for trainee managers direct from university, plays an important role. Many railways realised this during the period 1950–65, when statutory arrangements and financial restrictions did not allow such attractive salaries as those offered for careers in computer science or in the oil industry for example. The present situation seems much better[1].

364 *Training*

Professional training, initial and continuous, is a permanent concern of railway management. Very often railways do it themselves at operations and supervisory levels and they have developed a wide range of schemes, apprenticeship schools, inservice training and refresher courses. However one problem is posed by some fields. Most specialist techniques, both commercial and administrative, that are necessary to the railway are taught in universities or training colleges. But there is nothing comparable, save for a few exceptional cases, to careers in production, which are truly unique to the railway and whose rapid development and recent scientific orientation has been noted in this chapter. The railways must therefore carry out this training themselves. In some countries however there is an official teaching programme leading to careers in transport.

Normally, executives pursue their careers within one of the main administrative divisions (see chapter 5). But the increasingly multidisciplinary nature of problems and of future management jobs have led some railways to create their own institutions to provide an education at university level in all aspects of the railway. Some of these are open to executives from other railways. The development of international macrosystems also creates much greater geographical and economic openings. Some intergovernmental bodies, universities or railway organisations (in particular the UIC) have established courses devoted to international transport problems. The role of business journals, magazines and railway literature – the bibliography at the end of the book illustrates their considerable development – should not be underestimated, so long as the internal circulation of these publications is guaranteed.

It is an indisputable fact that since the war there has been a lack of openmindedness towards executive training, on a number of nationalised railways, and this has often resulted in very defensive attitudes. This opens the way to accusations of corporatism, coming from opponents to the railway and from governments. Such attitudes, contrary to the objective of public service to the community, are really self-destructive. The injection of new blood in the areas of marketing, economics and public relations and the development of common technology in control and computing have fortunately led to a reversal of this tendency.

The training of supervisors and managers poses a particularly acute problem for railways in developing countries, whose needs are too small to justify a local organisation. The most generally used formula in recent years has been to turn to the big railways, which take on trainees and integrate them into their own training programmes. This is definitely not the best solution for it takes the trainees away from their normal environment and may lead them to solutions which are not appropriate to the local problem. However a new, more realistic policy is being designed to organise such training on the spot but within an international framework, as we shall see on some macrosystems. Some railways in developing countries have even made the radical decision to entrust, for a limited period, the responsibility of their management to a team of specialists from a foreign railway. A plan as daring as this raises delicate problems of adaption to local conditions and relations with the government, social partners and with the staff of the enterprise. It has been widely used, with success it seems, in the much newer area of air transport, and it will be interesting to follow the results of the first experiments on the railway.

Finally, on railways in Latin countries, the post of director general is usually given to a senior executive, who has had an initial training as an engineer; in Anglo-Saxon countries, on the other hand, it is more usually given to lawyers or economists.

365 *Utilisation*

The running of a railway requires two categories of staff: sedentary staff and train crew. Sedentary staff deals with all activities not involving the accompaniment of trains. It is much the larger category and its organisation does not pose any particular problems. Part of this staff takes care of the continuous running of the service (safety, station service and depot service etc.).

Train crew consists of staff who drive engines or who are at the service of passengers. There are precise rules for the length of working hours and of rest time at home or away (the latter normally spent in internal railway hostels). For a regular service, staff organisation is determined, as for stock, according to the number of journeys done. In order to cover both regular and optional traffic, it is necessary to keep on call a reserve supply of staff, some of whom must be immediately available at home.

366 *Management–staff relations*

In all countries, the legislation concerning industrial relations has some provision for staff representation according to the various echelons of management, and agreed negotiation procedures in case of disputes. As we shall see in chapter 5, the present period of rapid transformation of the railway means that human communication must be further developed.

Railway workers' unions are powerful organisations, sometimes among the most important in a country, and their role in the running of the railway should not be underestimated. Their structure varies from a single union (sometimes with obligatory membership) to a number of independent specialist unions. The legality of the right to strike depends on the type of staff status. The history of railways is charged with memories of general strikes which nearly paralysed the national economy in the past days of the railway's monopoly.

37 Production costs

371 *Definitions*

Here I shall examine the costs of running a railway whose essential factors of production – lines, stock and staff – are stable, i.e. not at saturation point. The cost structure is complex because the railway produces a wide range of linked services using the same installations or the same stock in variable proportions.

The theory of railway costs, especially within the framework of marginalist economic theory, has greatly developed since 1950 and gives rise to lively controversies, especially in its implications for fare structures. I refer the reader to specialist publications, particularly those of R. Hutter and J. P. Baumgartner, and restrict myself here to the essential ideas.

First, general average costs and specific costs. The general average cost is the unit cost of traffic provision, or of the running of one category of train, on all or part of a railway. There exists then a whole family of general average costs, according to the railway's activities. A knowledge of general average costs is indispensable in order to establish the general balancesheet for the running of the railway by categories of service. This is one of the essential tools of management control.

The specific cost is the cost of a particular provision, all of whose characteristics are defined in detail. The calculation of specific costs form the basis of studies of fare structure as well as studies of investment, productivity and capacity.

Second, marginal costs and total costs. The marginal cost is the cost incurred in providing an extra unit of service. It is the change in costs corresponding to a change in volume of service supplied, divided by the change in service supplied. Feeding into the marginal cost is on the one hand expenditure directly linked to supply (energy, consumable materials, some of the staff and track maintenance), and on the other hand expenses which increase with the volume of the traffic but which once incurred must be covered over a certain period whatever the volume of supply (essentially crew and rolling stock). The marginal costs should establish an absolute tariff floor.

The total cost includes all the expenditure incurred in a particular provision. Its calculation is largely artificial, for it presupposes a breakdown of non-marginal expenses (management, line maintenance, financial and real estate taxes, etc.). The total costs can only be established as guidelines useful for pricing.

372 *Methods of calculation*

The calculation of costs presupposes the existence of an accounting procedure and a body of detailed statistics, and its accuracy depends on their structure. At present there is no universally accepted methodology in this field. For more than half a century US railways have been using a unified method defined by the ICC and which has just been revised. As for the UIC, it has been developing a method since 1950 which is recommended to all its members and subject to continuous improvements. The United Nations Economic Commission for Europe has brought out a method which is very similar to that of the UIC.

Some expenses can be directly ascribed via accounting and statistical data. For general expenses, distribution 'keys' are used, which bring in statistical data or inclusive rules. The different methods outlined above give precise rules for this. It is important to check the balance of the calculations, i.e. the agreement between the sum of costs and the sum of expenses of production on the network. Most methods need a final correction, starting from a balancing coefficient.

Some railways prefer, especially for the calculation of specific costs, to apply the standard pricing method based on the *a priori* determination of the normal volume of each production factor involved in supply. This method also demands careful calculation.

The practical execution of calculations is a very complicated job, which must be given to specialists placed in the office of general management and completely independent of accountants and statisticians.

373 *Structure and variation of production costs*

The results of calculation show that most costs are fixed and largely correspond to expenses which are not directly linked to traffic dispatch. There is also a much smaller variable part which does correspond to dispatch. Average costs diminish when production increases. As this concerns a quantity (passengers and tonnage) to be moved, average costs diminish when the volume of traffic increases and distance remains constant, or when the distance increases at constant volume of traffic, or finally when volume and distance both increase.

The railway thus comes into the category of enterprise which have 'growing efficiency', i.e. where the quotient of units produced per unit of production factors increases with production volume. In such enterprises, the marginal cost is lower than the average cost. It is important to stress that growing efficiency is not only the result of better distribution of infrastructure capacity. It also relates to expenses linked to dispatch. This point had been contested during the 'marginalist dispute' of the Sixties. Its validity was clearly shown by a very detailed study made by some of the Western European railways of the UIC. The growth of efficiency in dispatch, although lower than that of the infrastructure, is still significant. The consequences of the law of growing efficiency are essential, as we shall see in chapter 5, for the strategic management of the railway.

38 Operational management

This involves optimising the running of a continuous service for passengers and goods within a tactical unit of real time – the period of the timetable – over several months, but it often also involves working within periods of a few tens of seconds.

On railways where the whole service is homogeneous (metropolitan subways, mining lines) management is effectively reduced to the control of the execution of the traffic program , and to setting these procedures in motion as quickly as possible. There is sufficient advance notice to sort out problems because the number of possible cases is limited. 'Operations rooms' equipped with display units and run by controllers linked to all strategic points by a telecommunications network make it possible to have permanent control and feedback, as well as in some cases centralised traffic control. The length of technical hitches may then be considerably reduced.

But in the general case – a grid railway used by passenger traffic and random goods traffic – the dimensions of the problem change. Before the advent of telecommunications, railways could only check afterwards that the correct procedures ruling every detail of the different types of dispatch had been carried out. Since then it has become possible to ensure management in real time by means of 'management information systems' (MIS), feeding information corresponding to the principal factors of production into a telecommunication network and into data banks. I shall examine successively the problems of passenger and goods dispatch, and fixed installations.

381 *Management of passenger dispatch*

The basic service being made up of regular trains, management is reduced to controlling the utilisation and punctuality of trains, the use of rolling stock and to procedures for running extra trains. The control of train utilisation has been greatly improved by the development of centralised seat reservation, which means that train composition can be modified and relief trains provided in advance. The use of the stock of coaches can be followed individually because their number is limited.

Punctuality is controlled in real time on the lines themselves, as we shall see later on. The regularity coefficients are very often calculated based on a maximum of a five minute delay for metropolitan and suburban trains and 15 minutes for intercity trains. These now exceed 95 % on railways with highest speeds. 'The faster my trains the more punctual they are', said one railway director quite correctly. The other criteria of quality of service – comfort, catering, timetables, and interfaces – depend on sample surveys.

382 *Management of goods dispatch*

The main task is to ensure a regularity of transport within the framework of the timetable and procedures governing the details of freight dispatch (distribution of empty wagons, the criteria of tonnage of optional trains, their priority, diversion routes, etc.). Some railways are beginning to use MIS, making it possible to centralise the control, and sometimes the management, of local operations. The main systems now in service (TOPS, TRACS, GCTM, etc.) try to combine commercial functions and operation, and to make the best use of engines and rolling stock. One of their objectives is to be able to give advance warning to destination points of the arrival of wagons within a range of a few hours. Data banks are necessary to the running of MIS, for they deal with traffic as well as rolling stock; they provide the best way to optimise the next tactical period and to plan rolling stock maintenance. Up till now the volume and permanent distribution of goods stock could only be managed after its use and maintenance rules were based on time periods.

383 *Management of lines*

On all important lines, permanent traffic control is carried out from regulation centres which see to the execution of the train diagram. These centres are linked by telephone to stations and signal boxes and, on a growing number of railways, by radio to locomotive drivers. This means that measures to avoid accidents can be taken, relief trains slotted in, and full use made of two-way track signalling. We have seen that, where it exists, centralised traffic control combines traffic management and safety operations.

Observing the priority of trains when something goes wrong poses a difficult problem where there are tracks used in common by several lines close to big towns. Recent computing devices, in particular display units linked to electronic calculators, allow the controller, if warned sufficiently in advance of probable concentrations of traffic, to test several solutions. He can then choose the best and, by means of special signals, give orders to train drivers to slow down or speed up, so as to avoid stops in the middle of the line. In the case of tracks used both by suburban and long distance trains, such control zones may extend over a hundred kilometres. These are now in the experimental stage.

The adoption of an operational unit of real time of about ten seconds makes it necessary to have very sensitive and accurate time measurements. Time bases and electronic synchronisation circuits are used.

384 *Station management*

Large scale train diagrams called 'track occupation diagrams' control the normal use of large passenger stations, marshalling yards and the manoeuvres of the regular train service. Local management adapts these theoretical programs according to delays and extra trains. This is the role of a controlling body, most often situated in the central signal box and serving also to provide information to users and operating staff.

The operational management of small stations is tending to disappear with the concentration of traffic, automatic signalling, and the development of centralised traffic control and regulation centres. On a large number of railways, local train services at small stations are carried out by train crew acting according to instructions from the regulation centre. Only some commercial activities require staff. There is however some surveillance to ensure the safety of lines which do not have automatic signalling: checking that trains are complete, detecting overheated axle boxes, giving traffic information about trains to control centres. When the distance between important stations is too great, these constraints may justify keeping staff, whose responsibilities are essentially informative, except of course as far as safety is concerned.

385 *Productivity ratios*

Although management in real time has progressed, this does not mean that post-operational control has been done away with. Such control is based on statistics and makes it possible to formulate productivity ratios and prepare for tactical and strategic management decisions. The different types include economic ratios, such as quotients of traffic provisions (pk or tk) and a production factor; and technical ratios, such as running operating provisions (trk, tkh, skp, tkp) and a production factor. The calculation of these ratios, influenced by time, poses difficult problems of measuring certain factors (number of real hours of work, energy consumption, power of engines, etc.). In addition, the level of most factors cannot be adapted to short term fluctuations in traffic, some equipment is common, etc. In spite of these difficulties, the railways have since the last war, made considerable improvements in productivity in the areas of staffing, stock and energy. This confirms the progress in technology and production methods described in this and the previous chapter. One of the most significant results is a reduction in manpower, a policy which has been regularly pursued on most railways for 30 years in spite of sizeable increases in traffic and a reduction in legal working hours. What is more, this was achieved without the basic infrastructures being modified in any significant way, unlike the other forms of transport.

If it is without real significance to compare the productivity ratios of two railways because of numerous disparities mentioned above, at the level of the macrosystems, on the other hand several trends can be discerned. This is why I feel that the table given below deserves a reasonable amount of consideration. It consolidates the long term improvement over the last 40 years in results and essential ratios recorded by the three principal railway macrosystems in the world[1].

Without going into details here, these figures illustrate the growing efficiency of the railway, as well as the relative burden of lines with low traffic and passenger services, as the analyses in chapter 6 will show.

39 **Areas of research**

As with technology, this chapter has highlighted the profound transformation in railway production methods and operational management that has occurred in just a few decades. Management is now based on the extremely detailed programming of the execution of elementary operations, simple and repetitive, but carried out by a very large number of people performing different tasks, who are also very widely spread geographically. It was natural that cybernetics should be introduced here, in two main ways:

(*a*) as an instrument of command and control of repetitive processes;

(*b*) as an instrument of study and operations research simulation to optimise the combined subsystems.

This sudden awareness of the vital contribution of cybernetics to the future of the railway was quickened by the initiative taken in 1963 by Louis Armand, then General Secretary of the UIC, in organising the first World Symposium of Railway Cybernetics, with the participation of big international railway companies and expert delegations from universities and industry. This symposium was followed by three others which made it possible to examine new ideas, research and designs, and exchange valuable knowledge. But there is still much progress to be made, particularly in the field of improving goods

Table 3.1

		Europe		USSR		US	
		1934	1974	1934	1974	1934	1974
Length of lines (thousands of km)		271	252[1]	83	138[1]	383	335
Goods traffic (thousand million tk)		179	619	181	3098	432	1247
Passenger traffic (thousand million pk)		138	358	71	306	29	17
Staff manpower (in millions)		2.7	3	2.1	2.1	1	0.5
Ratio $\frac{\text{ktu (tk + pk)}}{\text{manpower}}$	(10^3)	11.6	32.6	12	162.1	46.1	252.9
Ratio $\frac{\text{ktu (tk + pk)}}{\text{km of line}}$	(10^6)	1.1	3.3	2.5	24.7	1.2	3.8
Ratio $\frac{\text{manpower}}{\text{km of line}}$	(unit)	10	11.9	25	15.2	2.6	1.5

1. These figures take into account the fact that after the Second World War 12 000 km of lines, which had belonged to 8 European networks, were transferred to the USSR.

dispatch from the MIS to the MOS system of improving goods management.

Another very significant development concerns specialisation. Whether it concerns trains or, more recently, infrastructures, the continuing development of specialisation of both fixed and mobile instruments of railway production is one of the best ways of responding to the growth of traffic. The consideration being given to the 'corridor' notion, with or without the construction of new lines, seems to open attractive prospects for the future. Here again, cybernetics offers the valuable instrument of simulation techniques.

So, in a few decades, operational management on the railway has developed, at least on important railways, from the advanced craftsman stage to the scientific stage, with all the consequences which such a rapid development involves for people and organisations.

4

Sales

41 Objectives

The traditional objective of the sales subsystem is to optimise the adaptation of supply, the characteristics of which were defined in the previous chapter, to demand. But because it here concerns a service, the notion of demand is a complex one. It may come from the free choice of its users, from previous choices of governmental authorities, or, most often, from both according to the case in hand.

The definition of supply is therefore directly linked to the transport policy of the country in which the railway operates. This is an essential component of management, and as such merits fuller discussion in the next chapter. I shall confine myself here to saying that, whatever the national political and economic doctrine, the development of transport economics has led to the distinction of two definite sectors:

(*a*) the 'competitive' sector, where at least in theory, market forces apply;

(*b*) the 'public service' sector, where a society imposes a certain quality of supply, irrespective of profitability and competition by means of financial subsidies.

In both cases, there are a great variety of markets to which the railway is liable to respond with a specific type of supply.

The sales subsystem interacts with supply which can be substituted by other forms of transport, and more particularly with combined subsystems. Finally, sales require their own organisation. I shall begin by analysing the supply offered by other forms of transport and combined supply, then describe the different types of railway supply, both passengers and goods, corresponding to various types of demand, and conclude by examining tariffs and marketing, which are the main interfaces with the system of users.

42 Supply offered by other forms of transport

421 *Roads*

The road is the oldest form of land-based transport, but the discovery of the internal combustion engine and pneumatic tyre caused a radical change. Road transport, revitalised by the car, is the only form that can be used equally well for individual or collective transport, that can serve the whole community, and provide a door-to-door service. Road infrastructure can be improved progressively, from track to motorway. Long bridges, viaducts, etc. make it possible to cross mountains and coastal inlets. Automobiles, powered by petrol or diesel engines (in some cases, gas turbines), have benefited from considerable improvements in energy efficiency and reliability. However, they are still dependent on petrol; alternative sources of energy (electricity, alcohol, batteries) do not at the moment promise great developments.

Two parameters play an essential role: the loading gauge (generally smaller than the railway) and the maximum load per axle. The constraints on these parameters have given rise to lively controversies, but experiments carried out in some countries and on certain types of infrastructure have confirmed the laws of mechanics: surface wear increases very rapidly with the load per axle. Most countries have fixed the limit of axle load at 10 tonnes, which makes possible road trains of two vehicles with six axles, 15 m long and a total gross load of 38 tonnes, or a payload of around 22 tonnes. This formula makes it possible to dissociate the engine unit from the load-carrying unit. Like the railway, the road has developed its own specialised trucks.

As for passengers, the car rapidly established itself as the instrument for individual transport. There are now about 250 million cars in the world, with a maximum density of one car per two inhabitants in the US and parts of Western Europe. The taxi is an individual car for common use. For group travel, short distance bus routes were developed first to provide urban and suburban transport. With the almost total disappearance of tramways, buses usually have specialised corridors reserved for them in the street system. They have the advantage of stopping more often, more flexible routes, and almost door-to-door service. The medium and long distance coach, sometimes with air conditioning, toilets and couchettes, is being increasingly used, especially in countries with low population density, because it supplies a unit of transport with low capacity, about 50 passengers. On the motorway, commercial speeds of 70 km/h are obtainable. There are now about 3 million coaches throughout the world.

As for goods, the success of the truck was immediate. The truck is a unit of transport which offers average capacity and door-to-door service. At present there are about 75 million in the world, 40 % of which are in the US, carrying practically anything over all distances and to all places. Use continues to rise and it has not been affected by economic recession and increase in the cost of oil. However, door-to-door service is not always the same as direct transport. The diffusion of traffic and the increase in capacity of trucks have led to the development of truck depots which play the same role as railway marshalling yards, with the three phases of concentration, transit and distribution, characteristic of freight dispatch. There are many of these depots around big cities and agricultural food markets.

The economic balance-sheet for road transport is difficult to draw up. Construction, maintenance and safety depend on the government or local authorities; use depends on companies or private individuals. The car has become an indispensable object of consumerism. Fuel makes up 25 % of the running costs of trucks. In some countries, the working timetables of long-distance drivers exceed the limits set by legal standards, which affects safety and creates distortions when compared with other modes. Road transport is not increasing in efficiency. Infrastructure costs are supported by the community, which partially recovers them through tolls or taxes. On this controversial subject, it is generally admitted that in some countries the private car is overtaxed, the Treasury making use of the car like alcohol or tobacco, whereas taxes on trucks go nowhere near paying for the wear and tear of road surfaces for which they are largely responsible[1].

The small-scale structure and the versatility of most road enterprises make it very difficult to apply a standard accounting procedure and conduct detailed analysis of costs. Road transport is also subject to very different national and international regulations; in most countries, the public transport service requires a licence. Awareness of ecological problems has finally highlighted some unfavourable factors, such as amount of land used, pollution and safety, which involves large investments (signalling, police) and a general speed limit of around 100 km/h.

The harshness of the recent increases in the price of petrol products[1] brought into question for the first time the irresistible rise of road transport and illustrated the danger of being dependent on a mode of transport which relies on only one source of energy. The major industrial countries with high oil imports, faced with the strength of the car-owning electorate and haunted by the fear of crisis in their car industries if their economies run into difficulties, have had to make do, in the immediate future, with what seem like derisory measures. However it is doubtful whether they can long delay certain decisions in this respect (must we increase the oil bill, even in reduced proportions, while the railway generally has excess capacity?) and investment (must we continue to give priority to improving the road network of countries in which the capacity of transport infrastructure is totally saturated?). Some less developed countries have already turned this corner by considerably raising the price of petrol, by putting restrictions on road traffic and changing the priority of their infrastructures. The late 1980s will no doubt make it possible for us to find our own way in the present fog.

422 *Navigation*

Dependent on the horizontal plane, inland navigation of interest here is found mainly on rivers with non-torrential flow regimes whose use requires little detailed infrastructure (principally port installations). But such rivers serve only a very small part of the continents and may be subject to seasonal restraints (freezing, low flow levels). Some rivers too difficult to navigate have been canalised, and artificial water courses, i.e. canals, have been built. But passing obstacles, however small, requires huge equipment: locks, lifts or water slopes. In the case of the St. Lawrence Seaway, such equipment has made it possible for navigation to penetrate right into the heart of the North American continent.

The size of unit loads varies from a few hundred to a few thousand tonnes. The convoy technique, either hauled or pushed, is widely used. Mississippi convoys dispatch 50 000 tonnes and those of the Rhine or Danube 6000 tonnes. Barge-carrying ships have been constructed for intercontinental transport. The decisive advantages of navigation are its very low energy consumption (about 1 litre of oil per tonne over 1000 km with a pushed convoy of 5000 tonnes), and its low cost. But it is slow; commercial speed, with night running, is around 10 km/h. In addition, navigation rarely gives door-to-door service, although the construction of factories by the waterside removes the need for small local services. These considerations preclude transport of passengers and perishable goods of high value and low tonnage from the navigation market, except when the river is the only system of land-based transport. Navigation, then, usually competes with the complete train load market.

Like the road, the economic balance-sheet of navigation is difficult to establish. The cost of infrastructure is supported by the community, which usually recovers only a small part from its users. Besides this, some river installations are built to meet several objectives, whose price breakdown is impossible to make, such as navigation, flood prevention schemes, irrigation, production of electric energy. The running of the service is distributed among a large number of enterprises and small companies occupy an important place. Regulations vary greatly.

423 *Aviation*

The newest mode of transport, aviation, has in less than 50 years, superseded maritime navigation in the intercontinental passenger market and has acquired a growing share of continental transport on distances greater than 500 km.

The development of short and medium range aircraft has led to jets with 100–300 seats cruising at subsonic speeds (900 km/h), designs which seem to have stabilised because methods that have been examined to bring the aeroplane into the town (short take off and landing (STOL), vertical take off and landing (VTOL), helicopters) come up against problems of cost and environment (noise) that are not easily overcome. Long-range aircraft have up to 400 seats; some with up to 1000 seats are expected. It does not at present seem possible to use supersonic long-range aircraft on simple continental journeys because of their very high cost and their effect on the environment. Though there are about 12 000 commercial aeroplanes in the world, there are very few types, all in international use.

Airports are reaching gigantic proportions, taking them further and further away from city centres. The air passenger therefore has to make land-based journeys of up to 50 km before and after the air trip and their cumulative length, together with the duration of formalities, is sometimes longer than the direct air journey itself. Finally, for a direct air journey in a subsonic jet plane, the commercial speed between city centres ranges between 150 km/h for 400–500 km and 500 km/h for 2000 km.

Though aviation has made considerable progress in productivity in recent years (about 85 % between 1965 and 1976 for big international companies), it remains an expensive mode of transport. It is entirely dependent on oil as a fuel, whose effect on production costs has practically doubled in a few years and is now about 30 %. However the time saved remains a powerful factor in attracting custom and airlines are making a particular effort to develop the tourist market through charter flights which give a substantial reduction in production costs. Air transport thus competes for a growing part of the market in intercity travellers.

Freight transport by air, although developing rapidly, represents only an insignificant fraction of all transport and its potential market is very limited in volume.

The economic balance-sheet of air transport is difficult to draw up. Airports are managed by national or regional authorities, safety by governments. Taxes on the use of airport facilities are in general lower than their cost. The companies using them are very often (except in the US) national enterprises, which are sometimes subject to political constraints on international flights. These also have complex regulations. The differences in fare structure are considerable in various regions, depending on competition. Faced with the continued growth of the cost of air transport, there has been a tendency in the last few years to place more importance on the idea of public service. But the policy of superliberalism imposed in 1978 by the US, which dominates world air transport, has opened a new era, the consequences of which at the level of continental links cannot yet be determined, particularly in Europe. Outside the US however, this policy generally meets with lively resistance, supported by the recent rise in the price of aviation fuel.

424 *Pipelines*

Transport by pipelines developed very quickly in the present century, mainly of crude petroleum products and gas. The average annual throughput of a pipeline 0.86 m in diameter is about 25 million tonnes of crude oil. By laying more pipelines, a total output of 70 million tonnes a year on routes several thousand kilometres long can be obtained in any climate.

Recently attempts have been made to extend this system to the transport of powdered solids kept in suspension in a liquid. These slurry pipelines could give an annual throughput of several million tonnes, mixed with approximately the same tonnage of water or methanol. At present there are only a few short slurry pipelines but plans are being studied for distances longer than 1000 km; these would have very low costs. Finally, the use of pressurised pipelines is planned for the transport of containers loaded with various slurry products or perhaps gas. These would require, if there were no 'points', a double track of pipelines, so as to allow the return of empty containers. An experimental line is in use in the USSR.

For their principal traffic, crude oil, pipelines do not compete with the railway, although a recent study (Alaska–Canada–US) has shown that railways should be able to provide a service at the same price under certain conditions. On the other hand, pipelines for refined products and for solids do compete with complete train loads, and the project mentioned above shows that research is being actively carried out in this field[1].

425 *Non-conventional guided systems*

1960 saw the appearance of new ideas for land-based transport, which break away from all track–vehicle contact. Initially the advantages seem attractive: lightness of equipment, the possibility of very high speeds, absence of noise, easy maintenance and simple infrastructure. Three systems have been proposed: the air cushion was recently abandoned after a suburban link project was not followed up; magnetic lift, which is being researched in a few countries, and the pneumatic propulsion tube. The numerous technical problems still to be solved may well be overcome given the necessary money and time. It will then be necessary to prove operational reliability, which requires experimental programmes using several vehicles on circuits of a length appropriate to the proposed speed, including points and signalling, as well as the simulation of technical hitches[1]. But the decisive criterion will of course be economic. Is there a market for such systems, whose designers aim chiefly at high-speed passenger links, about 400 km/h, i.e. between those of the train and the aeroplane? Will the potential advantages be sufficient to compensate for the numerous drawbacks, such as incompatibility and cost of infrastructure, access to cities, effect on the environment? At 400 km/h air resistance becomes so great that the aeroplane is more economic. Even if we can envisage in the next decades a few limited applications (very high-traffic-density corridors or town–airport links), the plan for a continental network, which is sometimes suggested, is too similar to a land-based Concorde for governments to take on any responsibility for it. This potential mode of supply cannot today be reasonably considered as capable of influencing the market in the medium term, but has to its credit the same adventurous spirit that brought high speeds to the railway.

43 Combined supply

Combined supplies link the characteristics specific to different modes of transport so as to provide multimodal services, answering certain kinds of demand better than a single mode. This is a conception as old as the railway which, since its origin, transported stage-coach frames on flat wagons. The development of road, maritime and air transport, has led to the multiplication of intermodal subsystems, and to try to bring together the main characteristics at a global level, I shall describe

- (*a*) 'multimodal containers', which can be used successively with transport vehicles belonging to different modes;
- (*b*) 'bimodal subsystems', characterised by the transport of a vehicle belonging to one mode on a unit belonging to another mode. For the railways, only two subsystems of this type exist: rail–road and rail–water.

431 *Multimodal containers*

These belong to the family of parallelepipedic boxes that can be loaded on a wagon, a truck or a road trailer, a ship and in certain cases, an aeroplane. There are several categories of container.

Small and medium containers are most often from 1–3 m^3 in capacity and can be handled horizontally or vertically. These are the oldest, are very widely used and usually belong to the railways. Air containers, with trapezoid cross-section of light construction, do not seem suitable for intermodal use.

Large containers, developed in the last few decades, are of two types. Maritime containers are of very stolid construction and may be stacked. Their dimensions (length 20, 30 and 40 feet; width 8 feet; depth 8 and $8\frac{1}{2}$ feet, maximum mass 30 tonnes) were fixed by the International Standards Organisation (ISO) without any consultation with land-based modes of transport, which therefore do not all have a suitable transverse loading gauge. Containers for land-based transport only have the same dimensions but they have a lower stacking capacity. 'Mobile boxes' are also used, at least in Europe. These are detachable but non-stackable road vehicle superstructures up to 12.3 m in length and maximum mass of 33 tonnes.

Vertical handling of big containers and mobile boxes requires vast terminals furnished with powerful hoisting machines. At first these were limited to ports but now they are being built in large cities with rail, road and river services. Large containers and mobile boxes belong most often to maritime or road transport companies or to private individuals.

The development of large containers has transformed maritime transport and has led to a renewal of the fleet of non-specialised ships for bulk transport. From the railway's point of view the transport of containers on wagons (COFC) does not pose any problem. Normal flat wagons may be used. However, specialised wagons are being developed, notably the articulated type with two or three chassis, and in the US, pit-cars, making it possible to stack two big containers 8 feet high. The huge variations in size of containers, a result of lack of cooperation and of the existence of certain dominant interests, is the cause of operational difficulties, and ought to encourage more consideration of intermodal factors.

Lastly, related to multimodal containers is the pallet of standardised dimensions, 0.8 m or 1 m × 1.2 m, with a maximum load of 4 tonnes; there are also pallet-boxes. The pallet is a universal unit of loading, carried by fork-lift truck, whose low cost makes it possible to store goods for longer periods.

432 *The rail–road subsystem*

The most widespread technique is the transport of a road vehicle on a railway wagon. Although it is, a priori, irrational to transport by rail a road vehicle with its own motor, and normally its driver too, there are two applications:

(*a*) transport of private cars, and accompanying passengers, by means of open or closed wagons with one or two levels;

(*b*) transport of trucks over sections of the journey where road travel is difficult (mountainous or desert regions). This is the case of the 'moving road' across Switzerland or Australia. The loading gauge, particularly in Europe, may dictate the use of low-loader wagons, with wheels of very small diameter.

The most widely used type is the transport of road trailers on flat wagons (TOFC). When, as in Europe, the loading gauge does not allow the use of ordinary flat wagons, low-loader wagons must be used. These include the 'well-wagon', loaded vertically, and the 'Kangaroo' or Wippen wagon, loaded horizontally.

The reverse technique, transporting a railway wagon on a road trailer, is unusual. It can make it possible to serve a factory or a warehouse not otherwise linked to the railway, but the length of the wagon is limited and such road traffic comes up against a lot of environmental restrictions. Experiments have also been made with air cushion apparatus.

433 *The rail–water subsystem*

Since the beginnings of the railway, ferries have been used to transport wagons across wide rivers, lakes or inlets. There two main types.

Multi-purpose ferries (carrying railway coaches and wagons, automobiles, road coaches and trucks, pedestrian passengers) are ships of medium capacity. If they carry passengers, certain standards of safety and comfort are imposed. They are used throughout Europe for journeys of less than 100 km.

Specialised ferries for the transport of goods wagons are much more economical to build and operate because they are of lower classification than ships transporting passengers. Their size varies between single pontoons (New York, Istanbul) and those with several decks, carrying 100 wagons more than 1000 km. (US–Cuba in 1935, the Baltic, Black Sea, Caspian Sea, Canada–Alaska). The low cost of maritime navigation has led to a revival of interest in these craft, especially on tideless seas, where terminus installations are very simple. The road–water subsystem has very widely developed the use of analogous roll-on roll-off vessels (Ro-Ro) and recently, in the Mediterranean, of superbarges travelling over very long distances. Such rail–water subsystems may by analogy be called 'rail-on rail-off' (Ra-Ra) operations.

44 **Railway passenger supply**

Cities are the obvious source of passenger traffic. Two very different markets can be distinguished: intercity links and daily commuting, urban and suburban.

These two markets both involve volume traffic, though the second is economically limited by the short distances. There is still a market for local links on a few railways, but these are inherently unprofitable because they lack both volume and distance.

441 *Intercity links*

Scientific analysis of the intercity market has only recently been undertaken, using models of different types. These models are difficult to formulate and even more difficult to validate because of the large number of parameters particularly influenced by the subjective behaviour of users (value of time, etc.) and the characteristics of other modes of supply. Interesting results have been recorded however, among them the distinction of three main reasons for travelling.

The first is the business trip. This covers a wide range of clients, from VIPs to junior clerks, but all with the common characteristic of having the journey paid for by the employer and not by the user. The cost parameter, without being negligible, appears to be of secondary importance to the duration and comfort of the journey.

In less than 20 years, in countries with high time-value, aviation has practically taken the whole of this market away from the railway, with the exception of links from city centre to city centre completed in 2½–3 hours, for which the aeroplane can do no better with a direct flight; and on certain night links where comfort and the timetable are particularly important. Competition from the private car remains low because of the congestion of roads and towns, the fatigue of driving and the frequency of accidents. The 'value of time' parameter is thus essential, but it can be evaluated only in limited and uncertain fashion.

An additional market, though geographically limited largely to Western Europe, has recently developed with the migration of workers to some industrial countries with shortages of manpower. The cost factor is of primary importance, and competition from roads (coaches) and air (charter flights) is very keen for frequent journeys of 2000–3000 km.

This second main reason for travelling covers trips for personal convenience, by nature dispersed, and for which the cost parameter is most often predominant. The frequency of these journeys has a certain correlation with income level. Competition from the aeroplane is lower, except in large countries with a low population density, because of the difference in price, but competition from the coach is very strong, for these can be cheaper and run more often. The evaluation of the cost of using private cars is very subjective; they are often preferred to the train when two or three people travel together.

Finally there are tourist journeys, concentrated in certain periods. The main criteria of choice are the cost and length of the journey. Experience shows that, in countries with high time-value, the railway has practically lost the tourist trade for rail journeys which cannot be completed in less than 24 hours at a suitable standard of comfort. However, transporting the car on the same train as the passengers has given the railway a new card to play in this field. The private car especially, and the coach and the aeroplane for distances greater than 1500 km, have captured the greatest part of this traffic, which has been developing continuously since the last war. In countries with low time-value, tourist journeys by rail of 2–3 days are still important. Finally we should not forget that more than half the world's population does not at present take holidays, which gives an idea of the potential of the tourist market in relation to rises in the standard of living in developing countries. Very long journeys for religious reasons depend on the same criteria as those for tourist journeys.

In order to cope with the range of geographical and economic circumstances, the railway supplies four types of intercity trains:

(*a*) traditional day trains;
(*b*) high-speed day trains;
(*c*) night trains;
(*d*) long-distance trains.

The characteristics of each will now be described in succession.

Traditional day trains running between 7 a.m. and 11 p.m. are the backbone of the passenger service on railways in heavily-urbanised countries. The objectives are commercial speeds (around 100 km/h) and frequency of service. The most widely-used mode of supply is the block train; the split train, which is not so fast, is mostly confined to Europe.

The frequency of the daytime train service varies considerably according to the urban geography of the network and its size. The extremes are without doubt the Netherlands and developing countries. In the Netherlands, with 330 inhabitants per square kilometre and a maximum journey distance of 250 km, an intercity linear-grid service based on multiple-unit trains operates with an hourly or half-hourly frequency. By contrast, in developing countries there may be only one train a day over sections of 400–500 km.

Some day trains now provide transport for cars. A new frontier has just been crossed in this field. Every weekend in 1979, the DB, in collaboration with the Association of Automobile clubs of West Germany, ran a successful hourly service of trains carrying cars over a distance of 150 km and avoiding the outskirts of Munich. The two terminal areas are situated within easy access of the great motorway routes from Scandinavia and West Germany to Austria and Italy. This service provides new evidence of the vast potential of rail–road cooperation in the interests of the general community (safety, pollution reduction, lower consumption of petroleum products). Such a formula is bound to be developed, for there are numerous analogous situations throughout the world.

Day trains with high commercial speeds (around 200 km/h) constitute a mode of supply born in Japan with the Tokaido, and soon to be extended to Europe and the US (the North East Corridor), though in a very limited fashion. These trains provide a service from centre to centre in 2–2½ hours over distances of 400–500 km, which competes favourably with aeroplanes. The Japanese formula will in principle be adopted everywhere; multiple unit trains and rhythmic timetables[1]. The train links concerned could well have very large traffic, so the objective is ultimately to achieve a maximum interval of around half an hour, so that a timetable need not be consulted and even reserving a seat would no longer be necessary. This would amount to a railway 'shuttle'.

Night trains are an essential commercial trump-card of the railway because they combine movement and rest through supplying a moving hotel. They also have obvious economic advantages. The minimum distance is rarely less than 600 km, an appreciably higher figure than the average journey made by a passenger on most networks; for several hours the passenger does not require any service. The objectives are comfort and convenient timetabling.

These trains are normally composed of coaches with sleeping compartments, though they sometimes also offer seating places. Their speed, usually not as high as that of day trains, make it possible for their load to be increased and for peripheral services to be included (mail, cars). Their range varies between 500 and 1500 km; there is a market, even for relatively short distances, for services scheduled between 9 p.m. and 7 a.m. The success of certain recent services of up to 1500 km scheduled between 6 p.m. and 9 a.m. and running at commercial speeds of around 100 km/h on routes with high demand shows that it is also possible to compete on these distances with early morning aeroplanes, which rarely arrive before 10 a.m. The level of comfort (smoothness of running, restaurant service, etc.) and timetabling are of primary importance. In addition, the market for passengers with cars has opened a new field with night trains. At first limited to seasonal tourist trains, this mode of supply is becoming year-long on major routes. The increasing cost of petrol should favour this development.

Long-distance trains, with journeys that take more than a night and a morning, are an essential part of supply on railways in very large countries such as the USSR, the US, Canada, India and China. The longest is from Moscow to Vladivostock, 9300 km and at present takes seven days. Such long-distance trains promote coastal trade and sometimes tourism. However they are still the means of transport mostly used in countries where time-value is judged too low for regular use of aeroplanes, whose cost is much too high for the majority of users. The concern for punctuality and the demands of refuelling and maintenance do not normally allow very high commercial speeds (70–80 km/h). These trains are normally made up of coaches with couchettes, coaches with seating places, and restaurant cars. They have become rare on international macrosystems where demand is lower. 'Cruise trains' made up of very comfortable stock could see some more seasonal development.

The development of the market and the supply from other modes of transport has led to a certain amount of simplification in comfort standards. Except for a very few, no railway offers more than two classes of comfort, defined by the amount of space given to each passenger and the quality of the furnishings. Except for a few railways in Western Europe, there are no longer any trains reserved for first class only. It is an indisputable fact that first class travel on the railway is superior to first class air travel, while the comfort of second class railway travel is at least equal to that of the economy class in aeroplanes or the coach, notably because only the train provides the opportunity to move about freely during the journey. Something similar is being developed for sleeping facilities. Just as there are business hotels and tourist hotels, the railways are limiting their supply, which was previously very diversified, to two categories of sleeping facility: the bed compartment, offering a 'room' with 1, 2 or 3 places and fitted with toilet; and the couchette compartment, with dormitories of 4–6 places.

Catering facilities are an essential element of comfort on journeys several hours long, but their cost in manpower is high. The present tendency is to offer lower quality but simpler facilities such as: self-service counters, bars, trolley sales, etc., while limiting the traditional restaurant to a few high-standard trains. Restaurants at stations and on trains often belong to the same enterprise and this contributes to a more efficient service on many railways.

Coping with peak traffic periods poses a problem that is usually solved, as we saw in the previous chapter, by devising a fuller 'summer service'. The necessary supplementary stock may be usable, at an acceptable cost, to satisfy the peaks of the winter service. On the other hand, the great majority of railways do not cater for 'super-peak' traffic, any more than do other modes of transport. A few Western European railways do, however, keep old stock, often outdated, to cope with this traffic without modifying prices, and may even reduce goods traffic for a few days in order to satisfy the social behaviour patterns of some clients; the financial and social balance of such a policy is difficult to evaluate.

442 *Commuting*

The characteristics of the market for daily commuting from home to work are well known: massive concentration for a few hours morning and evening, the imperatives of punctuality, low price and, as far as possible, comfort. During the day there may also be business trips, particularly urban ones, and private journeys, but their volume is much lower. The duration of these journeys is rarely more than one hour.

The characteristic mode of supply is the block train, usually with one class and regular timetabling. There is usually a rush hour timetable, with a frequency of a few minutes, and a timetable for off-peak periods, during which the frequency may go down to every half an hour or hour. The guarantee of a seat can rarely be relied on during rush hours. All metropolitan and suburban lines are electrified, which improves commercial speed and reduces environmental damage. Linear-grid services and interfaces suitably equipped with escalators ensure a good local service to urban centres, in conjunction with surface transport. Car parks at suburban stations provide the interface between the train and the private car.

443 *Local links*

Local railway services to villages and small towns were built only in countries where in the 19th century it was felt desirable (very often for political reasons) to connect the entire country by railway, the only mode of transport that could open up the countryside before the appearance of the car. Large countries with low population densities, where the railway was introduced later, did not as a rule build local lines unless there was substantial goods traffic for them to carry. The idea of local links applies equally, of course, to small stations on main lines.

The development of the private car and the coach, which better serve small population centres, have caused this traffic practically to disappear. A few markets remain: school students, factory workers, the diffusion of intercity traffic. Apart from these, the great reliability of the railway service provides a safety link in regions which suffer harsh climatic conditions. For what it is worth, there is also an 'historic' market, which is inevitably declining, except in a few regions with no road links. The most common railway supply is the single-class railcar offering a few daily shuttle services. Their commercial speed is rarely more than 50 km/h.

45 Railway goods supply

The initial purpose of the railway was to transport raw materials needed by new and growing industries. This remains an essential market; railways are still being constructed with this sole purpose. But very soon the railway showed that it is suitable for transporting anything, from a letter to the reactor of a nuclear power station, weighing several hundred tonnes. Since the end of the railway monopoly, these markets are no longer of the same financial interest, either for the community or for industry, and the impact of the other modes of transport has also been very variable. In order to analyse current goods markets, I shall adopt the following classification:

(*a*) bulk loads, of a minimum of several hundred tonnes;
(*b*) unit transport, from several tonnes to several hundred tonnes;
(*c*) light transport.

451 *Transport of bulk loads*

This market has grown progressively and now covers a very large range of products:

(*a*) raw materials; coal, ores, construction materials;
(*b*) industrial products; semi-finished and finished metallurgical products, chemicals, petroleum products, new cars;
(*c*) Agricultural food products; fertilisers, cereals, oil products, fruit and vegetables, drinks;
(*d*) urban wastes.

These involve homogeneous transport, normally integrated into quite a rigid process of fabrication or commercialisation. The main circuits are port–factory; factory–factory; mine/quarry/factory–distribution centre; agricultural market–urban market/factory. The extremities of these circuits normally have a private siding. The products transported being generally of low value, the essential parameters are price and punctuality, speed playing an important role only for perishable goods.

The mode of supply for this market is the complete train load. The optimum tonnage of such trains depends on numerous parameters specific to each kind of load: the gradients on the route, characteristics of locomotives, duration of stops and waiting at terminal installations, daily frequency, duration of terminal operations, etc. Complete trains dispatch most of the goods traffic on a large number of railways, which make it important to develop the quality of service, especially punctuality, and running speed.

452 *Unit transport*

This covers all transport of individual loads whose weight or volume amounts to the minimum use of one wagon. It is an extremely diverse market as far as products are concerned, and journeys are very dispersed. Although it does not lend itself very well to an analysis of the field, the following main areas can be distinguished:

(*a*) supplying heavy industry with raw materials of low tonnage;
(*b*) supplying medium and light industry with raw materials, fuel and semi-finished products;
(*c*) supplying wholesale outlets from factories, distribution centres and agricultural markets;
(*d*) delivering especially heavy or bulky loads.

The corresponding mode of supply is dispatch by single wagon, whose intrinsic slowness has already been mentioned. Also the market for unit transports is dominated by the road, which provides door-to-door service in all cases. However, the railway retains some important advantages in this field:

(*a*) private sidings can also ensure door-to-door service;
(*b*) fast goods routing has high commercial speed;
(*c*) speed is not always the determining criterion, but rather the low cost of transport;
(*d*) guarantees against delay in dispatch can be given increasingly often thanks to new systems of centralised traffic management:
(*e*) the nature, size and weight of many finished products make road transport impossible because it is too expensive or too dangerous;
(*f*) the single wagon, especially if it belongs to the client, offers the possibility of storage at reasonable cost because it does not immobilise an engine;
(*g*) recent formulae such as the mini block train and the semi through train eliminate, for unit transport of a few hundred tonnes, the disadvantages of dispatch by single wagon.

It is in this field of unit transport that combined supply – large containers or rail–road techniques – have opened new prospects. A few railways thought they had seen the end of dispatch by single wagon, for which they planned to substitute a linear-grid local service with semi through trains carrying large containers. After more than ten years, the experiment is far from proving conclusive and railways continue to marshal single wagons loaded with containers. Even though a number of analyses predict the great development of combined supply by exclusively land-based transport – intercontinental transport being already effected by these techniques – particularly in view of the recent rise in the cost of energy, many problems remain.

A large number of goods cannot be containerised because of their nature, weight or size. Since the volume of demand does not always justify a block train, a single wagon remains the only solution.

The management of a sizeable stock of large containers, usually belonging to companies with intercontinental business, is difficult and their use is low, as storage areas at termini show. 'Mobile boxes', for shorter journeys, would be better utilised.

If all terminal operations were transferred to the road, there would be large increases in traffic around large cities, which is contrary to present moves to protect the environment. Besides, in respect of railway operations, even in this extreme case it would never be possible to completely eliminate individual shunting.

Finally, as long as the railway must agree to transport any quantity of any merchandise on any route, dispatch by single wagon will continue to be necessary.

It is then quite reasonable to predict that dispatch by single wagon will retain an important role even though its market share will go down. Most railways are investing considerable amounts of money in modernising marshalling yards and increasing the number of private sidings, which is the best proof of this conclusion. On railways where the two kinds of goods routing, fast and slow, co-exist, the development of mini block trains and semi through trains, which lower the volume of routing, should mean that railways use only one kind in the future, one that combines low costs with performance between that of slow and of fast routing. On very small railways, it is doubtless easier to predict that single-wagon dispatch will cease. But when such railways are part of a macrosystem, and when there is inter-railway traffic which cannot be put in containers, a minimum of single-wagon dispatch will still be necessary, though restricted if possible to one marshalling yard.

Finally, the real problem seems to be the search for the best balance between private sidings and combined supply techniques, with a view to reducing, and if possible eliminating altogether, local train services with low traffic or services over distances which are too great. A possible strategy is to restrict the building of private sidings to those with a guaranteed minimum volume of traffic, and to encourage or even impose combined supply on users located on small lines or at small stations where the local service is too slow or too costly. This balance depends on the size of the railway, its geographical density, and its system of marshalling yards.

453 *Light transport*

This market dates from days of the monopoly of the railway. It has generally been taken over by the road, or at least by combined transport, the terminal operations being performed by road services starting from parcel sorting centres, linked by collecting wagons which travel, if possible, at night, when the volume of traffic means that such wagons will be filled. Some railways also use passenger trains to supply 'express' dispatch for parcels of high value (spare parts).

A particular case is postal transport. The advantage of the train over the road or air is that letters can be sorted in transit in specially-equipped coaches. These are normally attached to passenger trains, but a reduction in the length of stops means that it is no longer possible, on some railways, to provide a good postal service. So some postal services have completely abandoned the railway. Others, on the other hand, are developing postal transport by rail on major routes, but by means of specialised trains that usually run at night. It should be noted that the telephone and telex have decreased the number of urgent letters, and that techniques of tele-typesetting have considerably reduced the volume of newspaper transport.

46 Tariffs

461 *Objectives*

The schedule of supply prices must meet several objectives. It must ensure equal treatment of different clients in the same situations. A consequence of the monopoly position of railways, such protection of the client has always been one of the government's preoccupations and usually involves publication of prices. It continues, although it has sometimes been lessened, to place the railway in a more restrictive position than the other modes of transport.

Above all, together with the quality of service, pricing is the essential instrument of the commercial policy of a railway. This policy, as we have already said, depends on the national transport policy. But whatever the particular country's economic doctrine, the objective usually aimed for is to maximise receipts by fixing each tariff at an optimum level, between a minimum linked to the cost of providing the service and a maximum dependent on the state of the market, competition from other suppliers, the quality of service, and possibly on the balance of the budget. Finally the tariff must be clear and easy to develop.

I shall briefly survey the history of pricing and then examine the determination of the price range, present principles of pricing and finally fare structures.

462 *History*

Railway pricing has developed from the notion of the value of use. All transportable products were distributed among a number of price classes defined essentially by reference to the maximum level supportable by demand. Quality was gradually introduced in tariff schedules by creating several regimes of speed (low or high speed for goods, express and stopping trains for passengers) matched to different fares. The value of use of the transport of an item of merchandise is difficult to estimate; in practice, tariffs were based on the most frequent correlation between the actual value of the goods and that of its transport. The value of the goods thus finally emerged as an acceptable and simply-applied criterion. However, the development of other modes of transport, financial deficits, the development of transport economics, and the refinement of the calculation of costs have led to the 'ad valorem' principle being modified. The principal developments concerned the calculated minimum tariffs, adaptation to the different modes of supply (in particular block trains and combined services), the classification of transportable goods according to how they should be transported. The present situation may be characterised as one in which the 'ad valorem' principle is framed by an economic minimum and a commercial maximum.

463 *Pricing in a liberal economy*

After about a century of experimental trial and error, the marginalist theory attempted to answer this question rationally, following the work of Allais, Boiteux, Hotelling and Hutter on the application of the optimum welfare theory of Pareto to transport economics. Here it suffices to recall Pareto's claim that 'general welfare may be expressed mathematically as the maximisation of social efficiency, a state in which the satisfaction of any individual cannot be increased without diminishing that of another'. We can see that this maximisation implies the sale of all goods at their marginal cost. In an enterprise of increasing efficiency such as the railway, where marginal cost is lower than average cost, this necessarily implies a budgetary deficit, which must be covered by a careful distribution of tolls if balancing the budget of the railway is an objective of national transport policy, which is usually the case. If the theoretical lower limit of the tariff is the marginal cost, this will be sufficient only in an exceptional case.

The fixing of the upper limit is, we have said, linked to the commercial circumstances, theory suggesting the rule of equality of tolls between substitutable modes of supply. Application of this rule presupposes uniform knowledge of costs of different modes, and that each mode takes into account the aggregate of its infrastructure costs. Such a situation, unfortunately, exists only in theory. However, some general principles now seem to be accepted by most governments, as we shall see in the following chapter.

The application of marginalist theory raises a lot of practical problems. Its legitimacy is sometimes contested, and some railways determine their fares from total costs, although this method leads to an arbitrary allocation of fixed charges for a specific service. Others isolate 'profit centres'. Whatever the method adopted, the problem of the distribution of fixed charges can only be ignored in the very particular cases of specialised railways (metropolitan/subways, mining lines) and the validity of some types of distribution will always be debatable.

464 *Principles of pricing*

Tariffs usually take into account at the start two essential factors in the decrease of costs: volume and distance of transport. As far as volume is concerned, the decrease is general, as much for passengers (groups of different types) as for goods (tonnage requirements).

As far as distance is concerned, the formulae are more complex. The importance of fixed costs and of simplicity leads to the adoption in certain cases of flat rates independent of distance or according to zones (metropolitan and suburban lines, parcels, etc.). But most tariffs are based on the number of kilometres travelled, sometimes increased by a fixed sum corresponding to terminal costs. On extensive railways, the decreasing kilometric scale is very widely used, though some very large railways abandon the decreasing scale beyond a certain distance. The notion of distance can be refined depending on the traffic parameters (gradients on the route, kind of traction, type of train, etc.). This gives the 'weighted distance' which is substituted for the real distance of each homogeneous section, a fictional distance calculated according to a balance of transporting costs. This formula, which expresses the optimum application of the fundamental principles of the railway, can also play an important part in the comparison of costs of alternative modes of supply. It is applicable only to extended networks whose various lines have quite different characteristics; to my knowledge, only one railway at present uses it, solely for goods transport and within a restricted range. Its application to passenger fares on very good lines could appreciably change the competitiveness of air or road transport in favour of the railway.

The type of goods continues to play an important role, as we have seen. However the best use of rolling stock may make tariffication per wagon preferable, regardless of the goods; such a formula is being developed for large containers.

The differentiation of tariffs according to speed is not really justified, for we have seen that speed is often a factor of efficiency. It has now disappeared from most schedules of passenger fares. Some railways have also eliminated the dual speed scheme for goods, while others have linked it to the type of goods.

On some railways, the cost of traffic peaks is beginning to be passed on to passengers, with different fares according to the period. These are mainly reduced prices during slack periods; no railway has, up to now, tried to increase charges during super-peaks.

Terminal operations often have forfeits added to the actual cost of transport. Investments made by some clients (construction of private sidings, purchase of wagons) are taken into account according to modes of operation, with the aim of optimising utilisation (notably in respect of the payload/weight of wagon coefficient).

465 *Fare structure*

Usually fares are presented to the client in the following way. For each individual service, transporting a passenger or a given type of merchandise, they comprise under the application of the above principle a 'general' reference tariff and a 'specific' tariff for each mode of supply, depending on their technical characteristics, the state of the market and the existence of alternative supply. For goods, these separate tariffs in practice apply to most of the traffic. To make calculations easier and allow periodic adjustments of fares depending on the development of the economy or governmental decisions, fares often make reference not to individual prices but to scales of prices grouped in families expressing variation with distance or tonnage. A change of tariff is limited then to the use of a new scale.

Documents relating to tariffs also include some basic information: lists of establishments open to each type of supply, a table of real or weighted distances, classifications of goods, allowing them to be regrouped around several thousand basic positions. Unlike road tariffs, railway fares remain difficult to deal with; sometimes clients hand the problem over to specialised companies.

47 Marketing

It has taken nearly a century, corresponding to the period of the railway monopoly, for the commercial services of the railways concerned with tariffication and accounting to open themselves up to marketing. Since then though, development has been rapid. To gain knowledge of the market, standard techniques are used. Consumer research and customer relations departments have been created for the main products and work in close cooperation with departments dealing with calculations of costs. Publicity and wide-ranging advertising campaigns are now launched and the areas of contact with the public modernised. Even on the railways of countries with centrally-planned economies, where theoretically marketing is unnecessary, it is nevertheless used for international traffic, which remains largely competitive. Departments responsible for sales have given the railway a new specialist image and are now much more closely involved in designing and equipping rolling stock and interfaces (stations, etc.). Timetables are now drawn up by mixed teams of production and sales staff.

The results are already appreciable, especially on railways which had marketing organisations when the recent economic crisis hit the main industrialised countries. This sudden awareness of new realities of transport marketing has also made it possible for railwaymen to appreciate the need to abandon traditional modes of supply and to develop forms of intermodal cooperation.

5

Management

Unlike the previous chapters, this one does not begin by setting out the general objectives. Indeed we shall see that the management of a railway is only partially free to choose its objectives; it is not unusual for several, sometimes contradictory, to be imposed on it. With a service as essential to the community as transport, the state cannot remain indifferent. The relationship between a railway and the government, which depends directly on the national transport policy, must therefore be the subject of a preliminary analysis so as to determine the range of possible objectives, at least in the present situation.

In the introduction I identified the three modes of railway management, their interactions and the corresponding units of real time. Operational management, which corresponds to production, was described in chapter 3 and I shall not return to it here. Instead I shall look at tactical management and strategic management, confining myself to clarifying them as they relate to the railway, and referring the reader to the enormous literature treating the subject in general terms. The internal organisation of the railways will be dealt with next, then a few remarks about who has the power to make decisions will conclude this chapter and my analysis of the running of a railway system.

51 National transport policy

511 *History*

Invented in a liberal economy strongly influenced in Europe by Saint-Simonianism, the railway has always been subject to government control. This was initially to guarantee equal treatment for those who used a system which it soon became apparent was a quasi-monopoly. Other economic, social, political and military constraints have progressively accentuated the role of the state. But we have seen that the railway's quasi-monopoly was ended after less than a century by the development of the other modern modes of transport, which gave a new dimension to land-based transport by introducing wide areas of choice. Since the end of the nineteenth century, in many countries problems arose in running a railway under free enterprise, requiring the state to intervene because of the financial repercussions. Finally from 1917 onwards, planned economies were introduced in a growing number of countries, radically modifying basic options. So either by necessity or on principle, states have thus been forced to evolve a national transport policy. The contribution of transport to the gross national product is far from insignificant[1].

512 *The two policy doctrines*

At present, national transport policies derive from two basic doctrines: free market or central planning. The fundamental principles of each may be outlined as follows.

A liberal transport policy is based on the operation of a free market. Clients have free choice, according to the relative cost and advantages of different types of transport. In practice, its application comes up against a large number of obstacles, not least of which is to take into consideration the cost to the community of the different substitutable modes of supply. Even in the US, the epitome of the free market, no railway assigns all the costs to its users alone. Such a policy is also difficult to apply in a period of growth, as the first approaches to the problem by the European Economic Commission testify, while as we shall see in chapter 7, it is also not capable of supplying viable guidelines during periods of stagnation or recession. Besides, particularly since the last war, more or less compulsory national planning has been implemented in an increasing number of liberal economies. At the same time, almost all railway companies have been transformed into national enterprises, while the running of other forms of transport continued as a general rule to depend on private enterprise. It is not surprising then that conflicts of principle have arisen, particularly in the field of alternative modes of supply and so of commercial policy.

A pragmatic solution has gradually been evolved with the introduction of the idea of 'public service', which recognises that some services cannot depend on the economics of the market, particularly if it is necessary to run a permanent service[1]. A good definition of this idea, among others, was given by the CGST at Berne in June 1978:

'We are dealing with a service that is in the general interest, where a transport enterprise is maintained to fulfil certain obligations written into the law, which have economic, political, social, cultural, and energy objectives for the development of the nation or the protection of the environment. Such a service also takes into account possible exceptional circumstances and the needs of general defence. It is a question of providing a service that no commercially managed enterprise could undertake without appropriate subsidy. Providing a service that is in the general interest may be realised at the following three levels:

(*a*) transport infrastructure;
(*b*) supply of service (timetables);
(*c*) prices (fares).'

From this comes the co-existence within the same railway enterprise, of two sectors with different, indeed contradictory, purposes, which are nevertheless largely interdependent at the level of infrastructures. The other modes of transport are all in their turn involved in this idea of public service, and most of them also recognise such a duality.

A planned policy is based on an economy which is itself centrally planned. A national plan determines the criteria for the ruling policy, the domains of the different modes of transport and fixes the rules for their utilisation. There are many degrees of variation between the compulsory planning of countries with socialist economies and the 'soft' planning of some countries with economies that officially are liberal.

There is thus some diversity in the objectives at present considered by the management of a railway company. A very few countries, principally the US and Canada, continue to keep to a liberal policy, at least for goods traffic and within the limits cited above. The objective of management is the maximisation of profit within the framework of capitalist enterprise. But very nearly all countries with liberal economies now give their national railway company a double objective, corresponding to the two sectors defined above. In countries with planned economies, finally, the objective of management is the execution of the section on 'railway transport' in the national plan. This also fixes the plans of users which are state enterprises, as are those of other modes of transport.

52 Railway–state relations

521 *Principles*

The nature of the relationship between the railway and the state is an essential fact of railway management. Some state intervention, independent of all economic doctrine, is universal: safety norms, requirements of national defence, social legislation, and more recently, ecology. But major differences appear as soon as we tackle the questions of commercial policy and of financial policy, which is closely linked to it.

The railways of countries with planned economies do not, strictly speaking, have a commercial policy, since their objective is the application of a governmental plan. In particular, there is no direct relationship between costs and fares, these being fixed within the framework of planning policy as a whole. There is however a competitive policy for international traffic. There is no financial policy at enterprise level, and planning of investments is effected within a national framework relating to all modes of transport.

On railways in countries with liberal economies, and taking into account the evolution described above, the public service sectors follow an analogous scheme. The railways apply national or regional programmes for operations and investment, and fares are fixed by the relevant authorities. But the state also intervenes in sectors deemed to derive from the market economy. Throughout the whole world, including the US, fares are subject to governmental control. Complete freedom to run a transport service is almost universally denied to the railways. The state also intervenes, under the name of what is sometimes called transport coordination, through regulations and taxation, at the level of different modes of transport. Finally, the commercial freedom of the railways is limited because so-called liberal states currently use the railway as an instrument of their economic policy, particularly in periods of stagnation and recession. The laws of the market are therefore falsified and some costs transferred from the user to the taxpayer. Such a situation inevitably leads, on many railways, to significant accounting 'deficits', on which opinion is divided as to which part is attributable to the management of the railway and which is attributable to the national transport policy.

522 *Standardisation of accounts*

The Western European railways were the most directly affected by the serious consequences of non-standardisation, not only in their management but also in the service to the community because some indispensable investments had become impossible to make. They decided to propose to their governments, if not a policy, which was not within their province, then at least a few guidelines which seemed to be based on some economic common sense. Their ideas, channelled via the UIC since 1950, have paved the way for the development of a formula for the standardisation of accounts, which aims to give an exact picture of the financial position of the enterprise as if it were managed in normal industrial and commercial fashion. The railways do not discuss the political options of the states, any more than the status of personnel, pension schemes, etc. But they demand that the accounts be presented after distortions coming from outside the management have been redressed. Standardisation of accounts was first accepted by the European Economic Community, then recommended by the European Conference of Ministers of Transport (see chapter 7). It is now widely used throughout the world.

A catalogue of accounting items which offer scope for standardisation has been drawn up. Worth noting here are:

(*a*) the obligation to run a service and to transport;
(*b*) fares imposed;
(*c*) social services (pensions, social security) outside the scope of the general scheme;
(*d*) taxes and specific duties;
(*e*) maintenance of the infrastructure;
(*f*) financing investments.

The standardisation of accounts is the first step in bringing together the aggregate costs to the community of the different modes of transport. Its application makes it possible to obtain a more accurate view of the real financial position of the railway, and the consequences for the taxpayer of certain political decisions. We can now appreciate its usefulness to the community on decisions of a strategic nature. The conception of standardisation of accounts is also witness to the constant reflection on economic matters by railway directors.

523 *Forms of state supervision*

The institutional form of railway–state relations derives from previous considerations. In countries with planned economies, the railway has either its own ministry (USSR, China) or more often is a branch of the Ministry of Transport, and closely linked to the national planning body. Integration at government level is total and, strictly speaking, a distinct railway enterprise does not exist.

In countries with liberal economies the present situation is characterised by the almost universal existence of a national railway enterprise. A legal document, normally approved by parliament, defines the relation between the country and this national enterprise, and the conditions of contract details its obligations. Whatever the legal variations, the state dictates policy by expedient of annual budgetary approval (which sometimes comes form parliament as well), price control, allocation of investment credits and statutes for personnel. The government planning body, if it exists, intervenes in respect of investment programmes. While official supervision of the railway is normally assumed by a transport ministry[1], real supervision is ultimately the responsibility of the economics or finance ministry. In a few countries with liberal economies, the legal framework of nationalised enterprises, allows some autonomy in theory (the Netherlands). In a few others, there is a Ministry of Railways (India) or else the nationalised enterprise is incorporated into the Ministry of Transport (Italy, Denmark). In the US the very powerful Interstate Commerce Commission (ICC) has for a long time been concerned with fares. There is also bankruptcy legislation designed to avoid the service being ended if a private enterprise operation gets into financial difficulties. This is another way of recognising the railway's 'public service' character.

Thus the range of supervision methods is very wide. But it should be noted that in all cases the institutional relationship between the state and the railway is much closer than that which exists for other modes of transport in direct competition with it, which only in exceptional cases and in limited sectors derived from nationalised enterprises. Here again the burden of its lost monopoly continues to weigh heavily on the railway.

53 Tactical management

As we saw in chapter 1, tactical management comes into play between production, which is continuous, and the strategic plan of enterprise. There is permanent feedback between these three sectors as illustrated in the management of goods transport (figure 8[1]). Analogous schemas could be devised for other fields of activity. In practice the unit of tactical real time is the year, the universal unit of enterprise management, which also takes into account seasonal constraints affecting certain fields of the market (agricultural production, holidays), and technical running of the service (train timetables, maintenance work, etc.). The instrument of tactical management is the

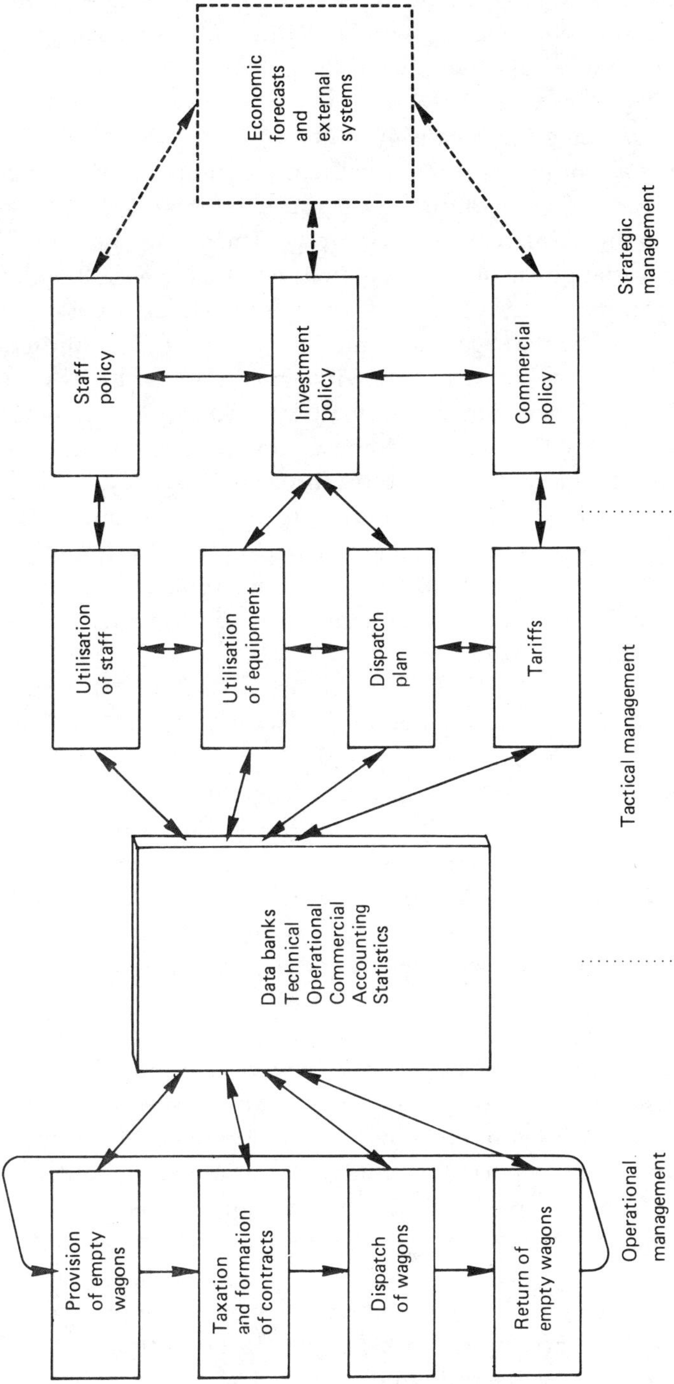

Fig. 8. The management of goods traffic

annual operations programme. The structure of its execution and control is essentially tied up with the internal organisation of the railway, which we shall examine later.

531 *The annual programme of operations*

This amounts to optimising the utilisation of the resources of the railway as a function of three principal elements.

The first is short term variations in traffic. The railway has long since passed the initial phase of rapid expansion (which was still occurring in aviation until recently). Except for new systems, or some very important development sectors, annual traffic variation is limited to a few percentage points more or less, with a maximum of 10. By virtue of the growing efficiency of the railway, upward variations can normally be absorbed by existing means; downward variations involve accepting underemployment of staff and resources.

The second element is the gradual realisation of an enterprise strategy, which as we shall see later, aims to improve the quality of the service while reducing its cost. The operations programme integrates the consequences for supply of annual stages of investment as well as internal reorganisation, etc.

The third element is the application of governmental directives, especially those to do with density of service, town and country planning, staffing, and taxation, possibly within the framework of a standardised accounting procedure. On the great majority of railways, it is necessary to draw up this programme in concert with the relevant government departments (transport and finance). As far as supply is concerned, cooperation with the users is also indispensable (programmed transport, passenger timetables, fares, etc.); it is normally effected through professional associations, chambers of commerce, etc.

532 *The operations budget*

This is the financial aspect of the programme. Forecasts of traffic, translated into production units (train journeys) are assessed according to category and modes of traction. Energy and running costs represent only a small part of the budget. The largest element, we saw in chapter 3, is made up of staff whose wages and duties are in practice determined by the government in a more or less direct way on almost every railway.

Operating costs are distributed in a very similar way on large networks which carry both passenger and goods traffic, now that steam traction has been eliminated.

Staff 60–70 %
Equipment and sundries 20–30 %
Energy 6–10 %

In the US, with only goods traffic, diesel traction and no guaranteed employment of staff, the figure for staff costs is only about 55 %, in spite of a very high percentage of accompanying staff on trains. The management's room for manoeuvre is thus reduced to modulating the rate of maintenance of rolling stock and track, a dangerous solution which has recently had catastrophic results in the medium term for the railways and regional economies in the eastern states of the US.

Operating revenues are divided up in very different ways depending on the railway. Apart from traffic receipts, which constitute the main part of the budget, there are some ancillary products corresponding to auxiliary activities.

The legal framework surrounding railway-state relations in countries with liberal economies very often dictates the annual budgetary balance. When standardisation of accounts is applied, the corresponding adjustments show whether costs have diminished or receipts have grown. The state has to compensate the carrier for insufficient fares. One may well agree with R. Bandeen that it would be more in keeping with the free choice of the user to transfer them to him, particularly for goods traffic. The same legal framework provides ways of covering up deficits, normally via the national budget, and their refunding. It also sometimes determines a compulsory contribution to investments. The approval of a budget depends on the relevant government department, and sometimes on parliament.

533 *Operations accounting and balance sheets*

These are generally established according to the national rules of accounting. The results are most often presented in such a way as to make it possible to appreciate, apart from the results of analytical accounting, the balance for the principal activities of the railway. The depreciation of the financial burden of stock renewal and investment is dealt with according to the regulations fixed by the railway–state conventions, which sometimes depart considerably from those in industry.

Most railways publish detailed annual reports which make it possible to appraise the management of the enterprise itself, as well as its performance as an instrument of the national transport policy. A very great effort has been made in recent years, with some success, to be as clear as possible, so as to give the public a full picture of the role of the railway as a service to the community.

534 *Productivity controls*

Annual productivity ratios make it possible, as we saw in chapter 3, to appraise the quality of the tactical management of the enterprise. It is possible to go further by applying techniques for calculating surpluses of total productivity. These techniques compare, from one year to the next, any variations in the volume of the different factors of production, which are then translated into values to render them comparable. The differences measure the 'surplus of global productivity', which can be distributed for the benefit of users and staff. These calculations are very complex, given the diversity of factors and services, but they enable better appraisal of the efficiency of the enterprise. They demonstrate in particular the economic results of the growing efficiency of the railway.

54 Strategic management

541 *Principles*

Strategic management takes a medium-term view of the future of the railway, that is to say, of its future place in the whole of transport supply. In spite of the great economic uncertainty at present, it seems possible to distinguish a few very generally accepted approaches.

In countries with planned economies, where generally there is not an aggregate surplus of transport infrastructures, the very important role of the railway is not questioned, even if road traffic, both collective and sometimes individual, and air transport are developing more rapidly[1]. The tendency is progressively to specialise the railway for the transport of massive loads over medium and long distances.

In industrialised countries with liberal economies, the economic crisis has shown that the financial difficulties of the railways are worsening considerably. This is a result of the law of increasing efficiency and it has also revealed a glut and aggregate under-employment of infrastructures. However there are still a few localised bottlenecks in both time and space. The high rate of growth recorded by most of these countries in recent decades has led to massive investments, essentially in motorways and airports. The user thus has a wide choice, in theory, but a choice distorted by lack of knowledge of real costs to the community and sometimes paid for dearly by the taxpayer. This waste of infrastructure is beginning to be noticed, and the increase in the price of oil that most of these countries import throws a new light on the cost of different modes. Nearly all the governments of these countries have undertaken research into the future of transport, particularly of the railway. The resolution of the 6th December 1977 of the Council of Ministers of the ECMT sums up their position quite well:

> 'The Council . . . believes:
>
> (*i*) that for economic and social reasons, the railway is an essential mode of transport;
>
> (*ii*) that the role of the railway should gradually be adapted so that it may better fulfil certain functions, which should be clearly defined both for the transport of goods and the transport of passengers, and related to one of the following categories of function:
>
> (*a*) the commercial function, solely for services which either are profitable or could become so: in this case the organisation of railways and their methods of operation shall have to be adapted to the changing market in both quantitative and qualitative aspects.
>
> (*b*) the public service function, to be limted to those services that the railway can provide in the best fashion for the community: in this case the government should pay a fair subsidy;
>
> (*iii*) that in all aspects of their operations the railways should be based on sound commercial principles;
>
> (*iv*) that the relations between the state and the railways should be based on a precise definition of the commercial function and of the public service function, and should respect the autonomy of railway companies, which in turn should be exercised within the limits of the directives issued by governments'.

This resolution is far from dealing with the whole of the problem; in particular it says nothing about the uniformity of calculations of cost, on the coordination of investments, or on energy factors. However, it does constitute, for its time, a positive attitude towards the future and makes official the two sectors of railway activity.

In developing countries an aggregate surplus of transport infrastructures is unusual. However this does not prevent the best roads from being built parallel to the most important rail links, especially from ports. The future of the railway is linked to minimum traffic and distance as well as to the need to open up the interior and develop international links.

542 *The two sectors*

The initial strategic task, for most important railways in liberal economies, is to define the boundary line between the competitive sector and the public service sector, a line which depends on the state and its transport policy.

For passenger traffic, this line is relatively easy to draw; urban and suburban transport unquestionably belong to the public service sector. There is also a tendency to extend this sector to all services of a regional character and even, in very large countries with low population densities (US, Canada), to intercity services, and so ultimately to the whole of railway passenger transport.

For goods traffic, the line is drawn differently. Most countries consider that goods traffic belongs to the competitive sector. However, as we have seen, an important part of the cost of dispatch by single wagon comes from terminal operations, while modes of combined supply offer good solutions. Defining the density of the railway network, a prerogative of the state, is thus inseparable from goods strategy.

The attitude of management towards the public service sector is by definition impartial and the accounts should be adjusted through standardisation. Medium term programmes and the corresponding financial investments are provided by the interested state departments.

For the competitive sector, by contrast, the strategic initiative of management is in principle complete. The objective is to maximise profit by increasing traffic and reducing cost, and investments are oriented to this end. At present, as we saw in chapter 4, the main strategic issues concern the forms and rates of development of combined supplies for unit transport of goods and their consequences for freight dispatch, the future of the transport of parcels, and the increase in speed, and the frequency and comfort of intercity passenger links.

I shall now examine in more detail the problem of the density of systems, then the main categories of railway investments, disinvestments, diversification strategies and finally the structures of strategic management.

543 *Systems density*

This is a problem specific to the railway, the only mode of transport (together with pipelines) to integrate infrastructure and operations. As noted in the introduction, the construction of new lines has seen a spectacular development in the last few decades, even in countries where there is something of a surplus of transport infrastructures, leading to lines being closed elsewhere. Is it possible, despite appearances to the contrary, to define a coherent strategy for this area?

All present strategies for the development of a railway system should arise from the following assumption: in 60 years the road has become a universal land-based mode of transport which is not hampered by geographical relief and which transports from door to door. It alone can serve a whole region, providing individual as well as collective transport. Road transport has, however, a number of very serious handicaps: efficiency is not increasing, the energy balance is low, it is totally dependent on oil, and pollution and heavy expropriation of land excludes it economically from some markets, as we saw in the last chapter.

This assumption was not valid for the nineteenth century, a period of rapid development for the railways in countries that were then the most industrialised. We may thus suppose that most of the present difficulties of the railway – I have emphasised their geographically limited character – arise from the fact that the automobile was invented a generation too late and we could and would not predict all its consequences. Indeed the railways most affected by financial difficulties (western Europe, eastern US, Japan) are precisely in those countries with the greatest excess of infrastructures, especially on local links where many small lines constructed at the end of the nineteenth century are still open, whose traffic has often been structurally insufficient to cover costs since its origin. Many lines were closed from about 1930 onwards[1], but there are still a large number with very low traffic in all these countries, carrying just a few thousand tonnes of goods per year or a few hundred passengers per day.

It was Dr Beeching, then chairman of British Rail, who first tackled this problem as a whole in presenting his famous plan of 1963, 'The Reshaping of British Railways'. The plan started from the observation that half of the system dispatched only 4 % of passenger traffic and 5 % of goods traffic, at a cost that was easily double the receipts. Dr Beeching proposed a substantial reduction in the length of the system, together with numerous other proposals concerning the concentration of supply. Very badly received at the time, this courageous programme was nevertheless largely carried out, and the results can now be seen. In 15 years the system was shortened from 28 600 to 18 000 km, a large number of stations were closed, staff was reduced from 475 000 to 248 000, and a very small working profit was made in 1977, on standardised accounts. From this experience, it is possible to measure the 'ladder effect', whereby large change becomes imperceptible if worked step by step.

In other Western European countries and in the US, governments or the railways themselves have instituted some similar plans. The ladder effect is particularly noticeable when, based on the main flows of present and predicted traffic, a system is designed *ex nihilo* just as it would be if it were built today, and were equipped and operated in modern fashion. Concentration of traffic, automation and telecommunication, the rational development of combined techniques and the reduction in weight of structures widely confirm the experience of BR. However, as soon as the public hears of such plans it has a tendency to react negatively and governments are reluctant to apply them in full. Such an attitude demonstrates how far its users still consider the railway as the sole guarantor of a permanent service, and for railway workers this is encouraging. But this view will not prevail given the new economic situation. The problem of network density thus has a political dimension and the example of Europe (which we shall examine in chapter 6 as the most typical) is not isolated. Certainly, standardisation of accounts should make local users think rather more carefully about this problem, for they are paying the cost. Standardisation of accounts is so far applied to goods traffic only in exceptional circumstances however, though for many railways the problem is situated there as we saw in chapter 4.

These considerations apply equally to developing countries. New lines, which will always be more or less parallel to main road links (except on specialised mining lines), must be limited to routes with heavy traffic and be restricted to serving important centres only. Conversely, some isolated systems with little traffic and short distances have already disappeared in favour of roads (Sierra Leone, Mauritius), and others are expected to suffer the same fate in Africa and Latin America.

544 *Investments in capacity*

Most important is the construction of new lines, which now must meet a few well-defined objectives.

(*a*) laying out a basic railway network in developing regions not yet or insufficiently served by rail.

This is the case in Siberia, the Middle East, the Sub-continent, part of Africa and Latin America. These are lines constructed after the road network, to which the preceding considerations are applied. A minimum traffic of about a million tonnes annually seems to be a reasonable size.

(*b*) crossing stretches of water (the English Channel, the Baltic, the Bosporus, the islands of Japan) connecting existing railways.
These special schemes are very often conceived as a combined rail and road link. They make it possible to avoid transfer from one mode of transport to another, and when they concern heavy traffic links (for example the Channel Tunnel) they could well bring an important part of the aviation market back to the railway.

(*c*) development of urban and suburban services: metropolitan and suburban lines.
Expensive to construct and to operate (tunnels, stations close together, large peaks of traffic), these lines are conceived as a public service. An interesting solution has been put into operation for the first time in Brussels. A surface network of trams was first constructed, then an intermediate stage called 'pre-metro', before the first real metropolitan lines. This formula may well be extended to other cities which have had the wisdom not to sacrifice their tram networks to the car too quickly.

(*d*) specialised lines to mines (in Asia, Africa and above all America). These show that the original purpose of the railway is still valid; their objective is purely commercial.

(*e*) finally, and this marks the debut of a new stage for the future of the railway, 'intelligent' doubling of heavy traffic routes that have reached saturation point (Japan, Europe).
These operations are normally conceived according to the 'corridor' technique described in chapter 3, and are distinguished by the degree of specialisation on the new line. They involve heavy investment whose cost per kilometre is difficult to compare when the great diversity of characteristics and the physical and human environment are taken into account. Such projects are subjected to far-reaching economic studies, using modern methods of evaluation (cost–benefit analysis, etc). The Paris–South East project proposed in 1970 by the SNCF was subjected to a positive second evaluation by the French Government and no doubt represents at present the most complete example of such a study.

Other investments in capacity have to do with the different stages of schemes to increase the traffic flows of lines and junctions examined in chapter 3. These are most often a response to a productivity objective, which will be examined later. Investments in rolling stock also combine the two objectives of higher flow and higher productivity by increasing the unit capacity of stock.

545 *Productivity investments*

These are essentially determined by the structure of costs, giving priority to reducing expenditure on staff and improvements in safety. Chapter 3 surveyed the main areas of recent development in this field and the considerable help cybernetics can bring in view of the fact that the railway is a guided form of transport. These investments cover four main areas:

(*a*) automation of functions depending on separate cybernetic minisystems which are used very frequently: control of signals, shunting of wagons, level crossing barriers, scheduling train crew rosters;

(*b*) concentration of dispersed installations by grouping them in installations with modern facilities, largely automated and linked to a telecommunications network; points, marshalling yards, seat reservation and charging centres, branches of administrative management (wages, stores, accounting, statistics, etc.);

(*c*) improvement in energy efficiency: programmes of dieselisation or electrification;

(*d*) a specialisation, as far as possible, of installations and rolling stock.

Maintenance techniques, both of track and rolling stock, are guided by the same principles and their industrial applications, while as we have also seen, increase in speed is often a factor in increasing productivity (Shinkansen, Paris–South East).

It is evident that most investments in productivity also increase capacity. It is the same in the opposite direction, except if the investment in capacity has as its sole objective the prevention of saturation. These subtle distinctions sometimes pose problems in allocating finance.

546 *Disinvestments*

The recent history of the railway has seen important disinvestments, demonstrating the need for a coherent policy.

When local lines, usually single-track, are closed, it usually means that the state regains the track bed and small separate buildings are sold. The most frequent problem arises when changes in traffic and productivity make it possible to reduce the number of tracks on a line.

Many lines with four tracks have been reduced to two, or to three with one having traffic in both directions, which is an effective formula. More controversial is the question of converting double-track line to single-track, which may arise when the track is due for renewal. Such an operation might prove profitable in the short term, even taking into account the fact that it is always costly to adapt signals; but it is not certain that it will be so in the long term. The regularity of utilisation will definitely suffer, and should traffic increase, non-recoverable investments will have to be made again. Some lines or sections of lines in mountainous regions were electrified during the time of steam traction for purely technical reasons. Again, the development of diesel traction sometimes justifies abandoning electric traction and its fixed installations.

Closing local services makes possible interesting disinvestments at junction stations (reduction of the number of platforms and technical workshops, simplification of signals). Extreme examples occur in the US, where monumental stations have been replaced by very small ones sufficient to cope with a few trains per day. Very valuable land, including that occupied by steam locomotive depots and construction workshops, can thus be recovered. Many railways develop a national policy for such matters, which are both remunerative and attractive to the clientele.

Finally, it suffices here to recall the importance of disinvestment in personnel, whose legal position was outlined in chapter 3.

547 *Diversification policy*

Since the beginning of the railway the companies adopted a policy of diversification whenever their legal position allowed it.

The first areas of diversification had to do with the idea of 'total passenger service': station hotels, travel agencies, road–rail tourist routes, etc. This idea was later developed to include other modes of transport, and some railway networks have rejuvenated their modal interfaces, as we saw in chapter 4. In many cases, the railway itself replaced local rail services with a road service; sometimes even inter-city coaches are used.

The development of goods transport by road has led many railways to create subsidiary road companies, particularly for parcels, warehousing and combined transport. Mixed rail–road firms have multiplied. Also, some railway enterprises have interests in other transport enterprises, maritime, air, or pipeline.

The increasing value of property and on some railways the need for staff accommodation have led to the establishment of railway estate agencies. Some railways have set up subsidiary companies for the construction and sale of rolling stock. Finally, since the last war and as a consequence of the establishment of financing institutions, some large European, Asian and American systems have entered the field of international railway consultation.

A few examples will serve to illustrate this policy of diversification. Extreme cases are found in the US, where some private railway companies make up only one part of a financial holding. A typical example is Illinois Central Industries of Chicago, which controls the Illinois Central Gulf Railroad among 20 other activities as diverse as Pepsi-Cola, Abex Corporation, insurance companies and financial institutions. South African Railways also control South African Airways, together with the main ports and road services. L'Agence Transcongolaise des Communications of the People's Republic of the Congo operates the Chemin de Fer Congo-Océan, the port of Pointe-Noire, and the navigation and ports of the Congo river basin (Zaire). The SNCF owns more than 50% of 11 subsidiary companies in the fields of transport (road, air, refrigeration), hydro-electric power, advertising, together with estate agencies and numerous organisations concerned with tourism, hotels, special transport, etc. British Rail possesses subsidiary companies for rolling stock construction workshops, ferries, ports, hotels. In nationalised railways, the diversification policy is obviously subject to government direction.

548 *Structure of strategic management*

State intervention is very significant, as we have seen. In countries with planned economies, the strategic rhythm is that of five-year plans, drawn up by a specialised body and including a section on the railway. In the great majority of countries with liberal economies, there is an institute of national planning, with more or less power, which integrates the transport strategy, while transport suppliers and users are normally called upon to participate in their turn in its work. While an outline transport policy may sometimes be prepared in this way, it should be noted that coordination of transport investment rarely gets beyond the planning stage.

In order to play an effective role in this dialogue with the state, the railways have created internal departments for planning and forecasting. In the absence of a clearly defined policy and in the present climate of economic uncertainty, these departments must often formulate their own brief. Leaving aside the all too rare large-scale projects, such as new lines, schemes that 'go with the current', and so avoid the simmering question of the density of the network, have the greatest chance of acceptance. In a few Western European countries, railways have started to draw up 'enterprise contracts' of several years duration. It is unusual for first proposals to be carried through without substantial modifications, but the idea appears to be in the interests of the community and deserves to be developed, insofar as such contracts will not be limited in the future to just the railway. But the supervision to which the railway has been subjected since its origin seems to offer such advantages to governments that it is right to ask if, beyond declarations of intent, governments do not sometimes fear a dialogue with a railway that has become profitable again.

Financing investments also come into the sphere of strategic management. The methods are various. Several railways and government agreements impose self-financing of some investments in productivity, especially when these are combined with renewing outmoded equipment. But as a general rule, investments are financed by a special budget fed from different sources: contributions or advances from the state, loans from banks or the public. As for rolling stock, leasing schemes have been developed to a certain extent for several decades. Indeed, these have led, as we shall see in chapter 6, to the creation in Europe of specialised finance associations. Finally, on many railways, users or groups of users build some wagons for themselves to ensure their transportation needs are met, and may offer them for hire.

55 Internal organisation

551 *Principles*

The parameters influencing the internal organisation of a railway are continuity of operations; coordination of this with the management of an infrastructure used concurrently to supply two totally independent markets, passengers and goods; the large number and geographical dispersion of installations; and individual staff discipline in the matter of safety.

In all railways there is a general staff, which is essentially concerned with tactical management but usually includes a strategic cell and operational units for geographical areas. The military vocabulary is not accidental, for there is some similarity between railway and military organisation. Some railways have indeed adopted a military structure, and retain the compulsory uniforms, which were the rule at the beginning of the railway. The size of the network in countries where the railway enterprise is the largest employer of manpower also poses the problem of an intermediate regional level of control, and the delegation of power to it. The cybernetic control circuit – transmission of orders, execution, monitoring and possible correction – makes available, through telecommunications, a very effective instrument to improve operational efficiency. But human relations remain essential.

552 *General staff*

The general staff has a double function. It is responsible for executing the programme of operations and the budget, and therefore the organisation of human resources and production equipment; and for working out strategic plans. These functions involve upward links with government departments, and downward links with operational units. In almost all railways in countries with liberal economies, the general staff has two levels: a board of directors and a general administration.

The powers of the board of directors varies between the extremes of an executive and a consultative body. It often includes representatives from the government, economic organisations and staff. Its province covers general strategy, commercial policy, financial problems and large investments. Though the government may not have a statutory majority on the board, it may nevertheless impose decisions made by the relevant ministries. On private railways, the board has the traditional form of joint-stock companies, its president being also able to hold the position of director general.

The general administration executive is made up either of a director general assisted by deputies, or a collective leadership. The various administrative tasks are distributed among functional departments responsible to the director general or the collective. The present tendency is to group these departments into large functional subsystems: sales, operations, maintenance of fixed installations and rolling stock, staff, computing, accounts and finance, supplies, real estate, and legal affairs. The formulation of the principal management guidelines (economic forecasts, statistics, calculation of costs), is often allocated to a specialised department. Finally there is, as already mentioned, generally a strategic department (research, planning). All this is in keeping with the development of methods of enterprise management. The multidisciplinary nature of many problems leads to the creation of interdepartmental committees. Some railways combine the production and sales departments, while others separate passenger sales from goods sales. Public relations, whose importance railways have sometimes realised too late, are usually dealt with at a higher level, while advertising meant for the client normally comes under sales. Liaison within macrosystems will be examined in the following chapter.

On railways in planned economies, we have seen that the notion of enterprise is not clear. It is the Minister of Transport or of railways who exercises authority. The responsibility for running the railway is generally given to a deputy minister, who sometimes takes the title of director general. Non-operational functions are most often carried out by other deputy ministers at an intermodal level, and I have underlined the major role of the central planning body.

553 *Operational units*

Railway production is, as chapter 3 has shown, almost totally programmed, with a unit of real time that may be less than a minute, and relies more and more on elaborate automation. The essential function of operational units is therefore the rigorous application of execution plans for each change in service, which determine in minute detail all actions that affect safety. We return here again to the analogy with military plans of operation. As soon as it is necessary to depart from the programme, either in normal service to cope with fluctuations in traffic, or in case of traffic accidents, the decisions to be made have, of necessity, regional consequences and optimising them can only be effected at a higher management level. This is one of the functions of the control relays I shall examine further on. The initiative of the head of the operational unit is thus limited to dealing with fluctuations of a purely local nature (for example, the assignment of platforms at a passenger station).

In spite of the development of automation, the size of certain units (marshalling yards, stations in large cities, depots and centres of rolling stock maintenance) demand labour forces which may comprise more than 1000 people. They are therefore usually given autonomous management. Small specialised units (stations, signal boxes, maintenance teams, etc.) are either consolidated by geographical zones, or attached to a large neighbouring unit, particularly for the management of problems to do with staff and rolling stock. Delegation of power in these fields (especially the management of operating staff) may be very extensive. On the other hand, it is unusual for operational units to have room for manoeuvre in the matter of fares.

If most operational units are specialised, the tendency to amalgamate them, accelerated by telecommunications, accentuates the multi-disciplinary nature of some of them (notably in the case of stations), which creates problems in training managerial staff. Some local installations which are not used very much may be able to do without resident staff, their functions being entirely carried out by crew on stopping trains.

554 *Control relays*

As soon as a network becomes more than a few hundred kilometres long and its manpower more than a few thousand people, control relays between the general administration and operational units become absolutely necessary. These must meet three main requirements.

The first, just mentioned, is operational. Continuous traffic surveillance and unusual decisions are consigned to multidisciplinary control centres, whose area of action may be more than 1000 km.

The second is commercial, to contact and canvass clients, and guide fare negotiations. Some railways, enjoying room to manoeuvre with fares, delegate part of this to their regional heads.

The third is administrative, to relieve the general administration of the responsibility for day-to-day management and so provide it with regionally consolidated information only.

Railways create regional administrations, relays between the general administration and the operational units, which integrate these different functions. The preceding analysis shows that their initiative is necessarily limited, except in the field of commercial activity. They may however play an important role in public relations (regional authorities, users associations, industrial and commercial authorities, etc.), and some railways have systematically installed them in the regional capitals of their countries. As in all management hierarchies, experience puts the optimum number of regional administrations at around ten and even the largest railways rarely have more than this number, unless they are themselves macrosystems (USSR). On very large railways, other control relays are also sometimes necessary between regional administrations and operational units, but they are usually monodisciplinary and cannot delegate power.

555 *Flow of information*

Railway management is largely dependent on information, most of which loses all significance after a few minutes. After more than a century of being snowed under by paper, often received too late to make a decision, which led to the railways being referred to as 'administrations', not 'enterprises', even during the time of private ownership railwaymen first looked at the development of computers with some trepidation. However they soon understood the operational and economic revolution that computers would bring to them as to all large enterprises, and soon made up for their initial tardiness compared to other sectors. Today, after several generations of computers and with the appearance of teleprocessing, the central nervous system of railway information has taken on a new form.

Chronologically, the first stage corresponded to the occasional use of computers sufficient for tasks of tactical management. This stage ended with the establishment, which is not yet complete even on large railways, of data banks that give instant knowledge and enable continuous review of the state of different production factors (staff, rolling stock, inventories of spare parts) and results of operations (accounting, statistics). This has had two consequences.

First, the establishment of information processing centres leads to economies of scale in producing weekly or monthly reports and accounts (pay, national insurance benefits, inventories, etc.). This has made possible economies in management and some decentralisation.

Second, it has made it possible to extract from this mass of data only the information needed by the different levels of control. These involve either statistical accounting syntheses or the automatic highlighting of anomalies. The well-known malady of mass computing has not yet been completely overcome; it is of a psychological nature, and computers in any case always produce far too many figures of no interest, more than anyone has time to read. Be this as it may, the march is in progress and the development of mini-computers should soon initiate a new phase of intelligent use of computers. In the meantime, teleprocessing has also influenced operational sectors, notably the management of goods traffic and the utilisation of passenger trains (seat reservations) through the linking of terminals and data banks. This field of application, we saw in chapter 3, is still far from being fully exploited.

However information flow cannot be reduced to the coldness of figures. An enterprise based on manpower and whose policy is partly imposed from outside cannot function harmoniously without a permanent current of human communications. Modifying a dispatch method, for example, cannot be done simply with one memorandum; the reasons and consequences must be explained. At the present time, when the railway is in rapid change and where an essential objective for the future is the reduction of manpower, staff information and motivation take on great importance. There are, as we saw in chapter

3, legal structures of communication; it is necessary that they play their role fully, but even this is not enough. On large railways especially, personal contacts and review conferences take place only at the highest levels, so that information received at lower levels tends to be distorted. Company newspapers may play a very useful role, for example through interviews with directors. Some experiments with televised talks given by the director general to staff have been made; these can be an effective way of helping all the staff of a railway feel they can participate in the development of their network.

556 *Bipartite management*

Should the management of the infrastructure be separated from that of operations? I shall only say a few words on this new idea for remedying the financial problems of railways in some countries with liberal economies. In my opinion it fails to recognise that these problems are not structural in origin and that the railway does not involve the same operating methods as roads or aviation. The integration of railway management makes it possible to solve numerous conflicts of priority between operations and maintenance; conflicts which are caused by using the same infrastructure for totally different types of traffic, passengers and goods, and with objectives sometimes in conflict to each other, public service and commercial competitiveness. A station is served by a single rail network, while a port or airport is served by several shipping companies or airlines, and the road by a multitude of individual or collective users. Further, the integration of railway management makes it possible to gain a sound knowledge of global costs. If such integration were broken up, it would be necessary to establish costly structures for coordination and decision. The advantages of such structures are hard to see, while the disadvantages appear immediately. To repeat, a problem located at the level of national transport policy cannot be solved simply by changing the structure of one of the modes of transport.

56 Power of decision

In conclusion, what can I say about the power of decision of the person who, with the title Deputy Minister, President or Director General, directs an enterprise often classed among the most important in his country? The answer is contained in the preceding pages. With some exceptions, this is the position of the 'number one' of a railway enterprise in the 1980s:

(*a*) he is not master of two thirds of expenditure, because staff has the benefit of guaranteed employment;
(*b*) he is only partially in control of receipts from the competitive sector because fares are subject to governmental control;
(*c*) he does not have the right to modify the structure of his network;
(*d*) he does not control the volume, nor, sometimes, the choice among investments;
(*e*) he is nominated and can be dismissed by the government.

It would be more truthful to call him a 'managing director' in the Latin American sense of the term, for his responsibilities, although enormous, stop there. If the railway is a service organised under the form of a national enterprise, it is no doubt right that he should be so. But it would be more honest not to accept the common claim heard in some countries with liberal economies that the railway is a 'commercial and industrial enterprise enjoying management autonomy', when it remains simply an instrument of national economic policy. The *deus ex machina* is of course the Finance Minister. A few exceptions still exist, notably in North America, but they only prove the rule. However we shall see later on that fortunately constraints are being loosened in some countries.

All this does not prevent a phalanx of senior civil servants from taking on this role with enthusiasm and efficiency. It crowns the career of a railwayman, or any one who comes of his own free will from the private sector. For it is, as Raoul Dautry has said, 'a man's job'.

6

Macrosystems

Very quickly, particularly in Europe and the US, railway systems linked up with their neighbours and thus created the first railway macrosystems (RMS). Nowadays there are only very few isolated railways and macrosystems are the rule. Many things have resulted from this in all areas of railway operation, with some railways now deriving the major part of their activity from the fact that they belong to a macrosystem.

I shall devote this chapter first to a theoretical study of macrosystems, then to a short analysis of the most important ones today by examining their specific problems. There will be frequent reference to the various international railway organisations, whose activity directly concerns such systems, but these organisations will be dealt with in the following chapter. A few characteristic types of railway system will become apparent from this survey.

61 Definition and parameters

611 *Definition*

An RMS is a collection of railways physically linked to each other and which exchange traffic. These railways may belong either to the same country (national RMS), or to several countries (international RMS). National RMS are becoming rare now that nationalised railway enterprises are the norm. They are only found in countries whose railways are based either totally or partially in the private sector (US, Canada) or in individual federated states (Australia, Switzerland). However, in a large number of countries there are still, besides the national system, various secondary systems, which are private or administered by local authorities but whose traffic exchanges depend entirely on the national system. Several continental RMS are made up of connecting national RMS and an international RMS, and there is a certain hierarchy implied in the idea of an RMS.

The physical link must be understood in a broad sense, taking into account the existence of combined supply and of the continuity of the contract of transport on road or sea journeys. The large maritime container stretches the notion of the RMS to cover all the railways of the world.

612 *Principles of operation*

The running of an RMS implies a minimum of standardisation and a structure for coordination and decision-making.

Standardisation may be effected in different degrees, from simple compatibility to complete standardisation in all the main areas: technical, operational, commercial, computing and legal. I shall look at each in turn.

The coordination and decision structure may vary from simple agreements of limited scope, protocols for regional cooperation, to permanent integrated organisation.

613 *Technical standardisation*

This essentially concerns the four following parameters: loading gauge, track gauge, braking, and vehicle couplings.

Establishing an inter-railway loading gauge is obviously indispensable. Its dimensions are those of the most limited loading gauge of the railways concerned, with some exceptions (the British loading gauge in Europe, for example). Enlarging the loading gauge of an RMS poses a very long-term problem which can only gradually be solved on only a few routes.

Having just one track gauge is no longer a major restraint. Nevertheless RMS are most often characterised by one gauge, or at least one gauge in the majority.

Unification of braking and coupling systems is absolutely essential because it concerns safety. However it is possible to use, though only as an exceptional procedure, 'transition' coupling between different types.

These different parameters need to be made subject to compulsory standards, which generally means distinctive marking on the vehicles used on inter-railway traffic. The development of the standardisation of rolling stock and of some fixed equipment shows advantages at the level of RMS at least as great as at railway level, particularly for maintenance and utilisation.

614 *Operational standardisation*

This touches on a field where the specific character of the railways (size, topological structure, nature of traffic, etc.) plays a major role. However a certain number of procedures must be standardised.

A primary objective is to eliminate all operations not dictated by commercial needs at the borders of railways. This requires agreements, most often bilateral, on the composition (length and tonnage) and routing of inter-railway through trains. The old practice of constructing marshalling yards at borders, sometimes double ones, is a waste of capital investment and is a cause of delays in dispatch, except when the border station is itself an important traffic centre. It should be done away with in any modernisation programme.

The running of engines from one railway to a contiguous one, may require the use of multi-current electric locomotives and makes it desirable to have uniform safety rules for the driving crew. Further, in the case of international RMS bilateral agreements should be at government level to facilitate execution of administrative formalities when crossing borders (police, customs, health checks).

The free circulation of rolling stock suitable for inter-railway traffic is based on the payment of a fee, most often calculated by kilometre (passenger stock) or time (goods stock). The strict application of these rules leads to long journeys with empty wagons, so some groups of railways have created a common stock of wagons of a single type for standard use within the pool, and periodic aggregate compensation replaces individual fees.

615 *Commercial standardisation*

This idea must be made more precise. It is very unusual to meet similar markets at the level of a large RMS, even if it is national. There are in reality a large number of regional markets, corresponding either to products or to flows of traffic but concerning only a part of the system. Commercial standardisation aims to facilitate the development of these different markets. The most general procedure is to institute, as with shipping and aviation, regional conferences which carry out studies of the markets, of fares and of routes for the sectors of traffic concerned, possibly taking into account national regulations.

The most logical pricing formula is that of the direct tariff, which disregards borders between railways but integrates the different parameters of transport cost (for example, the weighting of distances). It is relatively simple to apply to bilateral traffic, but often comes up against difficulties as soon as one or several transit railways are concerned, which benefit from the absence of terminal charges. When there are no direct tariffs, the inter-railway tariffs are determined by adding internal fares at border stations, an uneconomic solution dating from the days of railway monopoly.

There are however some identifiable markets at the level of RMS such as passengers, students and foreign visitors, for which contractual formulae have been put into operation with great success. For goods, one can cite combined transport by containers or semi-trailers (TOFC, COFC) for which inter-railway pricing has also been put into operation.

In any case, a certain amount of standardisation of internal pricing structure of railways considerably facilitates the fixing of inter-railway prices and their utilisation by clients. A standard nomenclature for goods is necessary.

Inter-railway traffic depends on complicated accounting for the distribution of receipts. In the case of international RMS, the adoption of a single unit of account and the establishment of a compensation office considerably facilitate operations.

616 *Standardisation in data processing*

This concerns multiple exchanges of information and messages between the railways of an RMS. Teleprocessing, as we have seen, has caused a profound transformation of the situation in ten years. The objective is not to unify computing equipment of the different railways, but simply to allow them to exchange assimilable information, this poses two problems: the codification of data and the standardisation of formats and procedure for transmission of messages.

617 *Legal standardisation*

This concerns the law relating to transport in international RMS. It requires legislation ratified by all the countries concerned and any changes take a long time to effect.

618 *Forms of inter-railway cooperation*

Respecting previously defined norms constitutes the indispensable minimum for executing traffic on an RMS, but this alone does not ensure the best service for its clients. Closer forms of cooperation are therefore necessary, especially for international RMS whose integration is not envisaged. It is probable that such integration would not improve efficiency, for it would pose problems of management, both human and technical, which would be difficult to overcome and would run the risk of producing white elephants. However by keeping to well-chosen sectorial activities, experience shows that substantial results can be obtained.

The pioneers in this field, almost a hundred years ago, were the Pullman Company and the Compagnie Internationale des Wagons-Lits for sleeping berths and the Railway Express Agency (US) for the transport of packages. While these enterprises were driven out of existence for reasons which will be made clear later on, they nevertheless showed the way. Under various legal forms, they functioned as ordinary subsidiaries of railways, managing a well-defined market. The advantages were obvious: a single transporter, quality of service, a prestigious image, standardisation of rolling stock, etc. The formula was revived around 1950 within the framework of the European RMS for two kinds of specialist traffic: refrigerated transport and transport in large containers. The railways concerned created cooperative subsidiary companies, with an autonomous commercial policy in negotiations with member railways. The formula showed itself to be very efficient, in spite of conflicts which might arise, for example, on pricing.

Without going to the extent of creating common subsidiary companies, some RMS established groups offering a standard product, such as day-time passenger trains of great comfort and high speed, sleeping berths, or even all intercity passenger services.

In the operational field, I have mentioned earlier the pooled utilisation of rolling stock which may cover not only vehicles but also certain loading accessories (pallets in Europe) and spare parts for standardised equipment.

Another form of operational and commercial cooperation concerns metropolitan macrosystems, i.e. those limited to the perimeters of big cities. These were born about a century ago when it turned out to be economical to operate communally in cities served by several railways. The objective of these was to connect either passengers ('terminal railroads' in the US, serving 'union stations' with engineering depots and shunting locomotives) or goods ('belt railroads' in the US and 'ceintures' in Paris, running connecting lines and possibly marshalling yards, and ensuring traction for transit trains.). Although the development of markets and structures made most of these systems redundant, the formula is being revived for passengers, but this time with the objective of integrating different systems of public transport in a large city: tramways, metropolitan subways, suburban railways and buses. It is a matter of optimising combined supply with respect to users, tariffs, interfaces and investment. The physical connection of metropolitan and suburban networks imposes further technical standardisation; such RMS normally depend on local government.

A very important area, finally, is that of cooperation in the fields of standardisation and research, which was outlined in chapter 2.

It remains necessary to ensure global cooperation, even more so since problems are now generally multidisciplinary. This is normally carried out by professional inter-railway associations, whose objectives and methods of work are the subject of the following chapter on international organisations. The description of the various RMS highlights the diversity of situations and practical solutions.

The forms of inter-railway cooperation are manifold and the member railways of RMS have become aware of the absolute necessity of intensifying such cooperation in all possible fields, even if occasionally they still come up against professional or governmental bigotry.

619 *The principal macrosystems*

We are now in a position to analyse briefly the main macrosystems of the world, beginning with what may be called the 'tricontinental megasystem', whose geographical limits will be explained further on, and which itself amalgamates five macrosystems.

Table 6.1

Macrosystem	Area (10^6 km^2)	Population (10^6)	Length of line (10^3km)	Passenger traffic (10^9pk)	Goods traffic (10^9tk)
Europe	4.8	480	245	365	610
USSR	22.4	260	140	330	3430
Far East	14.6	1030	65	125	540
Middle East	4.2	140	22	20	20
Maghreb	3.0	40	8	3	7
Megasystem	49.0	1950	480	843	4607
Subcontinent	4.2	770	70	180	150
Japan	0.4	110	25	220	55
Australia	7.7	14	45	7	30
Southern Africa	8.6	100	40	40	75
North America	21.3	300	420	25	1500
South America	13.6	160	85	40	110
Total	104.8	3404	1165	1355	6527

Throughout this railway journey it will be helpful to keep in mind the table shown above, which collects the essential data for macrosystems in 1977–8. Although the statistical bases are not strictly homogeneous, this table represents about 90 % of present railway activity in the world (metropolitan subways, isolated lines and railways excluded). Its formulation was made possible by the publication in 1979 of the first railway statistics from China, whose impact is considerable. Those who are surprised by some of these figures may be assured that they illustrate the realities of the present situation of the railway in the world.

62 The tricontinental megasystem

621 *Geopolitical data*

The tricontinental megasystem (TMS) includes Europe, the USSR, the Far East (Mongolia, Korea, China, Vietnam), the Middle East (from Turkey to Iran and Egypt) and the Maghreb (Tunisia, Algeria, Morocco). It is mainly oriented east to west and includes the longest distance that can be covered by the same railway vehicle (with four changes in gauge), more than 17 000 km between Algeciras and Ho Chi Minh City (figure 9).

An important geographical fact about the TMS is the existence of stretches of water or inland seas necessitating the use of transporter ferries for railway vehicles. These are widely used in Europe (British Isles, Scandinavia, Italy), in the USSR (Caspian Sea) and in the Middle East (Turkey).

Such a huge system cannot be immune to political upheavals. Consequently, in the early 1980s railway traffic was interrupted at the following borders: North Korea – South Korea, Lebanon – Israel, Israel – Egypt (the treaty signed in Washington on 26 March 1979 envisaged the re-establishment of this link).

622 *Economic role*

One might be tempted to consider the TMS purely from the theoretical viewpoint of the geography of transport. But this would be wrong because for three quarters of a century it has played a considerable economic role.

In the field of passenger traffic, it is enough to remember the Trans-Siberian Railway, which provided, up to the age of aeroplanes, the quickest link between Europe, China and Japan, and the combination of the Simplon–Orient Express and the Taurus Express between London and Basra (for India) or Cairo, all this for a long time under the

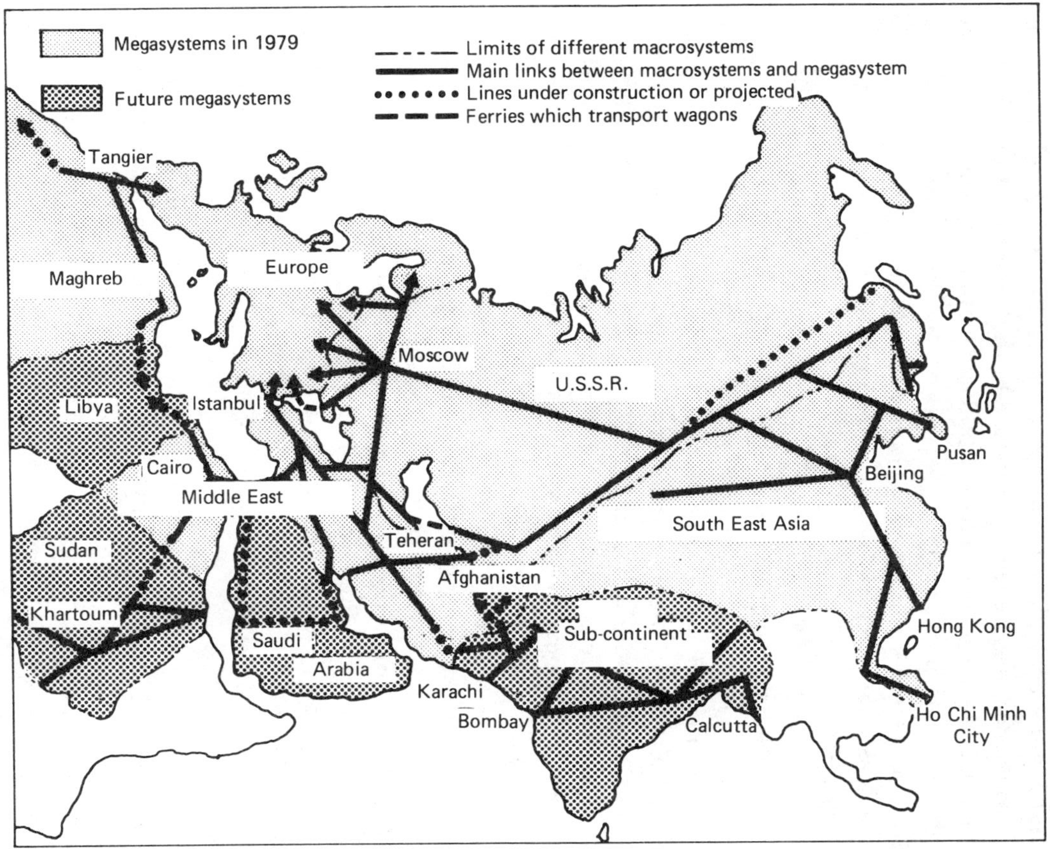

Fig. 9. The tricontinental megasystem (TMS) (internal grids not shown)

prestigious name of the Compagnie Internationale des Wagons-Lits. While air transport took their traditional clientele of diplomats and business men, most of these long-distance trains still run, with a different but much more numerous clientele, and new links have been created. A real spider's web, centred on Moscow, has been woven since 1950 by the Soviet railways, with block trains and sleeping coaches reaching almost all the European capitals, from Helsinki to Athens, as well as Teheran, Beijing and Pyong Yang in Asia.

But it is especially in the fields of goods traffic that rapid development has begun, notably between the RMS of Europe and of the USSR and Middle East. Block trains run between the USSR and numerous European centres, as well as between Europe and the Middle East. By contrast, the traffic between the USSR and China is very low. The large container has found a privileged place on the TMS, notably by providing a terrestrial bridge between Europe, the Middle East and Japan via the USSR, as well as between Europe, the Middle East and the Maghreb.

623 *State of standardisation*

Technical standardisation has reached a sufficient level to allow traffic exchanges.

The European loading gauge (called the 'universal UIC') is also applied in the Middle East and the Maghreb, but the USSR and the Far East have a more generous one. The result is that the rolling stock on these railways can run only on certain routes in Eastern Europe and the Middle East. In addition, Soviet railways have built a large fleet of rolling stock with the UIC loading gauge (especially coaches).

The track gauge is not uniform. Standard gauge is used in Europe, except Finland and the Iberian peninsula, in the Middle East, the Maghreb, in China, the Koreas and North Vietnam; the Russian gauge (1.524 m) in USSR, Finland and Mongolia; the Iberian gauge (1.676 m) in Spain and Portugal. Installations for changing bogies have been built at the main border stations and lines with the Russian gauge have been built in Czechoslovakia and Poland to serve large industrial complexes.

The air brake is used throughout the TMS. The only real technical constraint is the variation in couplings: manual in Europe, Middle East and the Maghreb, and automatic SA 3 in the USSR and Mongolia. American automatic couplings used in China, Korea and North Vietnam are not compatible with USSR couplings.

On the TMS operational standardisation is of secondary importance, for stops at border stations on the different RMS are long for numerous reasons not always to do with the railway. Through passenger trains must wait at border stations in the USSR for about two hours, and less than half of this time is needed to change bogies. There is no such problem between Europe and the Middle East.

At the commercial level, there are several inter-RMS conferences: the Balkans–Near East Tariff Union, the France–Spain–Portugal–Morocco Conference, MTT transit tariffs, etc.

At the legal level, links with the USSR and beyond are regulated by a more complex arrangement, for as we shall see there are two international conventions, one applied by the railways of liberal economies and the other by railways of planned economies, members of Comecon.

The unified codification (see chapter 7) of all data, necessary for standardisation in computing, was adopted without any difficulty through collaboration between the UIC and the OSZD.

624 *Prospects*

A decisive development in the geography of the TMS should be reached before the end of the century when the Sub-continent (Pakistan – India – Bangladesh) is connected to the Iranian system. A wagon will then be able to run over 11 000 km between London and Calcutta. A railway is planned for construction in Afghanistan and linked to Iran, the USSR and Pakistan, and just a few hundred kilometres of line to Xin-Jiang will be enough to open a third link between the USSR and China. The construction of the Libyan network is in progress; it will finally link the railways of the Mediterranean by connecting the Maghreb to the Middle East. There is also a link between Egypt and the Sudan and subsequently to East and Southern Africa. On the other hand, linking the TMS to the south east (Bangladesh–Burma, Thailand–Kampuchea–Vietnam), which would necessitate quite long sections of metric track, some of which was constructed for military reasons during the last war and later abandoned, does not appear to hold any economic interest, and this part of the 'Trans-Asiatic Railway Project' will probably never be constructed.

Little progress is to be expected in the field of standardisation. The principal obstacles are still the differences in loading gauge and coupling between Europe and the USSR, which see the most important inter-RMS exchanges. The fact that the railways of the TMS belong to two different economic systems and observe two different legal conventions, does not make the definition and implementation of a common commercial policy any easier.

Although with the connection of the Sub-continent the TMS will then serve two thirds of the world's population, it will not see the development of significant inter-RMS passenger traffic for a long time, with the exception of links between Europe and the USSR. It will very likely be the same for goods traffic. Sea routes via the Suez Canal compete effectively with land-based transport on links between Europe, the Gulf countries, the Sub-continent and China. It also avoids numerous border crossings. The nature and volume of future exchanges between the Middle East and the Sub-continent are difficult to predict. They will depend on the rate of industrialisation in the countries concerned; however, increased transport of many agricultural food products is likely. On the other hand, Europe–Middle East–Maghreb traffic should develop when huge wagon transporter ferries are put into service in the Mediterranean, operating over distances that may be as much as 2000 km.

63 Europe

The European macrosystem is the oldest and is still one of the most important in the world. It is also the most complex because of its international character; it comprises the railways of 25 countries (only the Albanian network is at present isolated). Its long history has been distinguished by numerous political and economic changes, without its fundamental unity ever being questioned, and it has had to face the most severe handicaps. It is worth examining at some length because it is exemplary in many respects.

631 *Handicaps*

First of all, geography does not favour the railway in a large part of Europe. The continent is practically cut in two, from the south west to the north east by a succession of mountains: the Pyrenees, the Alps and the Carpathians. Crossing the Alps in particular required, apart from the construction of long tunnels, the adoption of rigid layout characteristics which limit the capacity of the lines.

The islands and peninsulas around the margins cut off the British Isles and Scandinavia from direct land communication. Passenger traffic between the United Kingdom (12 % of the European population) especially London, the biggest city in Europe, and the continent has rapidly slipped from the railway, except in summer, to the advantage of aviation and combined road–sea operations. The same situation has developed to a lesser extent in Scandinavia. Wagon transporter ferries account for only an insignificant part of goods traffic.

The small area of many European countries means that internal transport distances are very low. It is possible to travel more than 1000 km on eight railways only.

The main European railway axes are all more than a century old. Their fundamental characteristics have remained the same, though traffic has never ceased to develop. Of course, substantial improvements have been carried out – electrification, track doubling, automation of signals – but the original layouts remain.

The dimensions of some basic parameters, dependent on both the geography and the age of the system, constitute another handicap. The 'universal' loading gauge does not allow the transport of lorries and road trailers on normal wagons because their height is too great, and it is necessary to build special wagons for heavy traffic use. On some routes an enlarged loading gauge has been introduced in response to the main needs of combined transport; this is a priority question still under study.

The length of storage sidings varies between 700–750 m which makes it impossible to exceed an average load of 2000 tonnes gross for ordinary goods trains and 5000 tonnes for block trains carrying bulk loads. This limit stems from the survival of manual coupling; the present financial situation has made it necessary to postpone the installation of automatic coupling to the end of the century.

From the economic point of view, the European railway system developed at the dawn of the Industrial Revolution, and so in an environment which was still largely rural. This explains the construction, especially in the Western and Central European countries, of numerous branch lines. The advent of the car and the spread of urbanisation quickly made them uneconomic, indeed redundant. The relative poverty of Europe in basic raw materials and the rise in the cost of their extraction have further reduced and sometimes completely halted the great flows of heavy traffic which gave rise to the railway. Now there are only two substantial generators of such traffic: coal in Poland and iron in Sweden. Heavy industry has moved to the ports and is served by sea.

Finally, political upheavals born of the last war have practically eliminated the international passenger traffic over medium distances which was once very important in Central Europe (Vienna–Budapest, links between Berlin and the other great German cities, etc.). However they have considerably developed the inter-railway traffic of countries with planned economies.

The most important handicap is without doubt institutional. The European RMS includes, apart from 25 national railways, a few regional and many secondary railways. The national railways are totally independent, even if some countries are part of the international amalgamations to be examined in the following chapter. Their length ranges between 200 and 35 000 km, the volume of their traffic varies in a ratio of 1 to 200 and their international role is extremely variable. They make up a veritable hotch potch[1].

632 *Quality of service*

For passengers, the quality offered in Europe is probably the highest in the world. An increasing number of trains travel at 200 km/h on existing lines and a speed of 160 km/h has become common on the main arteries. Intercity interval-service timetables are increasing. Night trains allow comfortable journeys of up to 1500 km and more and more of them carry cars. Constant improvements are being made to suburban services on all railways: frequency of service, electrification, double-deck coaches, and town–airport links are multiplying. The quality of service has significantly lowered, however, in two fields: catering and long-distance links, especially the Balkans. The prestigious names given to the old luxury international trains of the CIWL, which are sometimes retained, now disguise no more than a skeleton service of one or two through coaches coupled to a succession of national trains, which implies long stops for connections, much shunting and finally low comfort and very poor commercial speed, discouraging the residual market whose continuing existence is affirmed by the development of long-distance coach services. Finally, the European RMS is characterised by an extremely regular 'summer service' to the seaside and a 'winter sports service' constantly developing, to the snowfields.

For goods, block trains of 3000–5000 tonnes gross dispatch the major part of heavy traffic, and their field of operation extends to several markets. The mini block train of around 500 tonnes marks a breakthrough. The transport of perishable goods is an expanding market due to a network of specialised trains, some of which travel at up to 140 km/h . Slow goods routing offers slow speeds only, but the use of a huge network of private sidings and the development of schemes with guaranteed delivery help the railways to retain a large clientele. Finally, combined techniques are being developed quite rapidly.

633 *Standardisation and cooperation*

Essentially, these are pursued within the framework of the UIC (see chapter 7), of which Europe constitutes the central nucleus.

Technical standardisation is complete and has now reached the stage of complete standardisation of some fixed equipment (rails), of rolling stock, (goods wagon) and rolling stock components, (bogies, brakes, coupling). The maximum load per axle is 20 tonnes. Research is continuing to bring this up to 22 tonnes.

In the operational field, timetables and the composition of international passenger and goods trains are fixed every year by European conferences. The crossing of borders has made great progress; within countries of the EEC, a growing number of passenger and goods trains cross without stopping. Sleeping passengers are no longer systematically woken up by the police and customs except at a few borders, and the application of resolutions made at the Helsinki Conference should gradually put an end to this altogether, which is at present holding back the development of a considerable market. A system of international goods dispatch is being put into operation, which will make it possible to increase the number of through goods trains, without reshunting at borders. Finally, two rolling stock pools have been created, which amalgamate several hundred thousand wagons, and a similar pool for pallets now exists.

In the commercial field, the system of regional tariff unions has widely developed. A few passenger tariff unions have been created with success at a European level; Interrail for the youth market and Eurailpass for the extra-European market. A common nomenclature for goods was made compulsory in 1977.

Standard accounting has been achieved for the distribution of receipts. A central clearing house makes it possible to limit money transfers. The gold franc unit of account used at first had to be abandoned because of the instability of the international monetary system. It was replaced in 1976 by the 'UIC franc', calculated from a basket of European currencies and subject to adjustment. It is the first example of a unit of account applicable to many countries, not all of whom are members of the IMF and whose currencies are not convertible.

Data processing standardisation has been installed very quickly and without difficulty since 1960; it covers all fields concerning international traffic. Basic codifications are fixed, as we have seen, at the level of the TMS.

Legal standardisation was achieved a long time ago under the aegis of the OCTI (see chapter 7), which administers two international conventions, the CIM and the CIV. The countries of the CMEA have put two conventions into operation, the SMGS and the SMPS, which differ little from the CIM and the CIV and control transport law between their networks. However they apply CIM and CIV rules for transport involving other networks.

As a general rule, European railways made their primary objective to present themselves to international clients as a single carrier. Two major successes have already been achieved: the common subsidiary companies Interfrigo and Intercontainer. In the passenger field the Trans-Europ-Express Group offers the most demanding clientele on some networks a service which is similar to the pre-war Pullman trains of the CIWL. But one cannot move on without mentioning the considerable retreat from the principle of the single carrier resulting from the disappearance of this company, which for a century was the sole – but prestigious – instance of a truly 'European' railway. For many reasons, political, commercial and financial, a profound revision of the former situation was required. But what actually happened was effectively 'balkanisation'. Ten pools or subsidiary companies now share the sleeping coach service on the European continent, while that of couchette coaches has remained the domain of each railway, a situation without parallel anywhere else in the world. The Soviet railway has become the principal European international supplier of sleeping accommodation and practically the only one on most east–west links. This development is regrettable in a field where, as we saw in chapter 4, the railway mode of supply is particularly favourable and where the dimensions of the railways and the increase in speeds continue to accentuate the railway's international character of this market.

International cooperation is also developing among private organisations that are owners of specialist wagons; a good example is the Spanish company Transfesa. In the field of combined transport, the International Union of Combined Rail–Road Transport Companies (UIRR) has begun fruitful collaboration.

634 *Recent developments*

The situation of the railway in Europe, already difficult between the wars for reasons described in the introduction, was worsened by the Second World War. Used to the limit of its potential because it was vital for military operations, the railway benefited after the war from the absolute priority given to restoring it to good working order, thus making it possible to revitalise the community and to restart industry. Between 1950 and 1955, a working railway was recreated in Europe, but it was a 1938 model. One could not complain about this to railwaymen or to governments, for they had no other choice, but the situation rapidly turned against the railway. While rolling stock built according to pre-war standards between 1945 and 1950 has by now practically disappeared, the same is not true for fixed installations: the layout of the lines has remained unchanged and the large connecting stations, lines of secondary importance and steam locomotive sheds are now partly unused or in too great supply. The same period has seen the true birth of commercial aviation, the development of road transport with the construction of a motorway network, and the construction of large canals. Thus the technical standard of the equipment of the main railway routes has, by force of circumstances, been exceeded in quality and in capacity by the other modes.

On the other hand, the division of Europe into two political and economic blocks rapidly led to rather different patterns of development. Countries with planned economies,[1] convinced of the lower cost to society and the growing efficiency of the railway, and having in addition for a long time had a reserved attitude towards the private car, have resolutely backed the railway and given it priority in investment. The construction of motorways was begun only recently and at a rather slow rate, while commercial aviation, whose users are mostly civil servants, is relatively little developed. The natural navigable waterways play an important role in regions with gentle topography, but large canals have not been constructed. Railway traffic has therefore been regularly and strongly developed and now faces serious problems of capacity on the principal routes, notably those connecting with the USSR, the principal source of raw materials and the most important client of these countries (traffic exchanges at present reach around 100 million tonnes per year). A competitive situation on the other hand, has arisen between railways with planned economies, especially for traffic which serves Scandinavia, the Balkans and the Middle East.

Countries with liberal economies, by contrast, have given priority to road transport, which has become the symbol of individual freedom and of rapidly rising standards of living, besides being the creator of many jobs. Its expansion was made easier by political control of some sources of oil supply, which ensured that the price of fuel would be maintained at low levels. Road transport thus developed practically without hindrance and has successfully pursued its policy of 'creaming off' the market without any opposition, from social legislation for example. The same is true of aviation, which was stimulated by the US, then the only supplier of equipment, and which saw rapid expansion thanks to infrastructures of military origin. The value of time for the Western European businessman has come into alignment with that of his American counterpart. In about a decade, the railway thus lost a considerable part of its first-class passenger traffic for journeys of more than three hours, to the benefit of the aeroplane, and an important part of its goods traffic by unit loads, to the benefit of the truck. In spite of continual gains in productivity, the accounting 'deficit' of these railways has worsened wihout much hope of improvement, given the obligation to provide and operate a service and the controls on tariffs maintained without any appreciable relaxation. Many economists and politicians saw the future of the railway as limited to the urban and suburban transport of passengers and to the transport of bulk goods. No investment in capacity or radical improvement of quality of supply thus seemed justified and in fact remained exceptional. Railwaymen, powerless to oppose such views, had to witness the gradual degradation of an industry to which society had shown its affection and which had always shown itself to be the most economic in numerous fields and irreplaceable in the event of difficulties.

It was in the years 1965–70 that the reversal in perspective mentioned in the introduction took place. A more rational appreciation of the operating costs and the development of the different modes of transport, proof of the growing efficiency of railway administration during a period of industrial boom, the influence of ecology and the need to place before the user choices that are economically justified, have led countries with liberal economies to the doctrine of the double objective analysed in the previous chapter. The determination of true costs has made great progress thanks to the standardisation of accounts formulated by the UIC. The competitive passenger sector has been limited to intercity links, the obligation to provide and operate a service has been lessened on many networks, and greater flexibility in prices, sometimes in theory amounting to freedom, has been granted. Governments have become aware of the problems of capacity on main routes, and the launching of the first generation of new lines has provided spectacular evidence of this renaissance of the railway in 'liberal' Europe.

Without doubt, the consequences of the crisis in the European iron and steel industry and the rapid rise in the cost of oil products brought with them a sizeable fall in the goods traffic of railways in liberal economies, and shortly after slowed down the progress of railways in planned economies. But intercity passenger traffic, which has benefited everywhere from an enormous technical and commercial effort, shows positive trends. We are now witnessing, at last, the closing of the gap between the two types of development on the European networks, although the situation of the railway is still much more favourable in countries with planned economies. A few figures on the growth of railway traffic provide sufficient evidence of this (table 6.2).

In 1978, the Polish network represented around 10 % of the length of the European RMS but alone transported (thanks to coal trains) 25 % European goods traffic. This figure reaches 50 % when just two other railways (the French and the Czech) are added.

635 *Probable trends*

It is likely first of all that energy problems will in the short term impose some shift in the policy of priority given to the automobile in countries with liberal economies and a partial but substantial transfer to rail of unit transport over medium or long distance. Some governments have already understood this, for they now readily finance private sidings and intermodal techniques, and encourage railways to continue electrification. The recent imposition by several countries of a road tax on trucks, so as to make them bear part of the cost of wear and congestion of roads, is a movement in the same direction. The considerations discussed in chapter 4 in view of the latest increase in the cost of oil products are particularly applicable to Europe.

In other respects, the human geography of Europe can only reinforce the competitiveness of the railway in the passenger field. There are

Table 6.2

Traffic	Economic policy	1958	1968	1978
Goods	planned	100	167	239
(tk)	liberal	100	116	119
Passengers	planned	100	111	131
(pk)	liberal	100	98	122

more than 30 cities with more than 1 million inhabitants, which need urban and suburban railways (there are more than 30 metropolitan lines either working or under construction). More importantly, the distance between these cities is rarely more than 300–400 km. This puts the railway in an excellent position with regard to air transport. The impact of new high-speed lines will be decisive in this respect. Further, demographic stagnation (there will be a population increase of only 10 % from now to the year 2000, the lowest figure in the world) and the subsequent increase in the number of old people should work in the railway's favour. Safety, comfort, the absence of restrictions on rail travel for medical reasons, and cost constitute advantages of the rail over road and air transport. The time-value of a retired person is much less than that of someone still working.

Countries with planned economies can only be confirmed in their policy of priority to rail. As for countries with liberal economies, it is becoming urgent that they finally draw up a coherent transport policy. The liberalism extolled by the EEC (see chapter 7) but biased by an insufficient knowledge of fare costs is becoming impracticable and contrary to the interest of society; nor will it resist the pressure of tomorrow's energy situation. One can do no better than quote the recent proposals of M. Gscheidle, the West German Transport Minister:

> 'The guarantee of free choice of carrier also implies that its limits are to be fixed according to competition, when it bears on the existence of different modes of transport. There must be a European solution to take charge of costs on the main traffic routes. . . . We have an absolute need for a European rail policy' (Munich, October 1978).

The growing efficiency of the railway, the double sector theory and the standardisation of accounts have largely opened the way to the formulation of such a policy.

An essential aspect of this policy is the rejuvenation of the system. While its total length (250 000 km) has varied little in 30 years because of the completion of the basic lines in some countries (Spain, Finland, Yugoslavia), this is compensated by the closing of secondary lines, especially in France, Great Britain, Poland and Sweden. However, its utilisation has greatly changed. On many railways, half the lines are now sufficient to haul 80–90 % of the traffic. In the same time the road network, more than three million kilometres in total length, has built 25 000 km of motorways, a figure which increases every year, as does the number of automobiles, which now exceeds ten million vehicles.

This rejuvenation concerns both the trunk and the branches of the European railway tree. The first stage consists of continuing to prune dead or dying branches – around 40–50 000 km of lines, or nearly 20 % of the present network – concentrated chiefly in regions with no heavy industry and low population density. This raises political problems, and it is regrettable to see that an appraisal, in my view erroneous, of certain economic factors, including the increase in the price of oil products, and ecological factors, has led to a slowing in the rate of closures in the last few years. There is an obvious contradiction with the affirmations of liberalism here.

The second stage consists of re-invigorating the healthy trunk and branches, nearly 200 000 km, including 40 000 km of major international routes identified by the master plan for European infrastructure drawn up by the UIC (figure 10).

The modernisation of this system and the rate of increase in the capacity of the principal routes must be speeded up. In some cases, the solution rests in the optimisation of a corridor, with the eventual construction of new lines, naturally conceived for very high speeds (200–300 km/h). The UIC plan envisages about 6000 km of such lines; 1300 km are under construction and 400 km already in service (Italy and Poland). However this programme may well prove insufficient in the short term in some countries.

As noted in chapter 7, a major bottleneck occurs at alpine passes, where policy has often complicated the situation. The St. Gotthard, the Lötschberg and the Arlberg passes are situated wholly within one country, and so belong to a single railway; they can be crossed without formalities and their flow rate, even if the approach lines are partially single-track, is very high. This is not true for Mont Cenis or the Brenner passes, for example. Although the two approach lines have the same gradients their signalling and their type of electric traction are different, the weight of goods trains is unequal, and stops for customs and administrative operations are necessary. The throughput of these lines suffers because of this, although improvements are still possible in the coordination of operations and automatic coupling will dramatically increase the weight of trains. The slowness in making decisions contrasts with the rapidity of construction of motorways and alpine road tunnels, both national or international, whose number grows regularly.

One cannot pass on without mentioning the Channel Tunnel, conceived in an intermodal perspective, whose coming into service

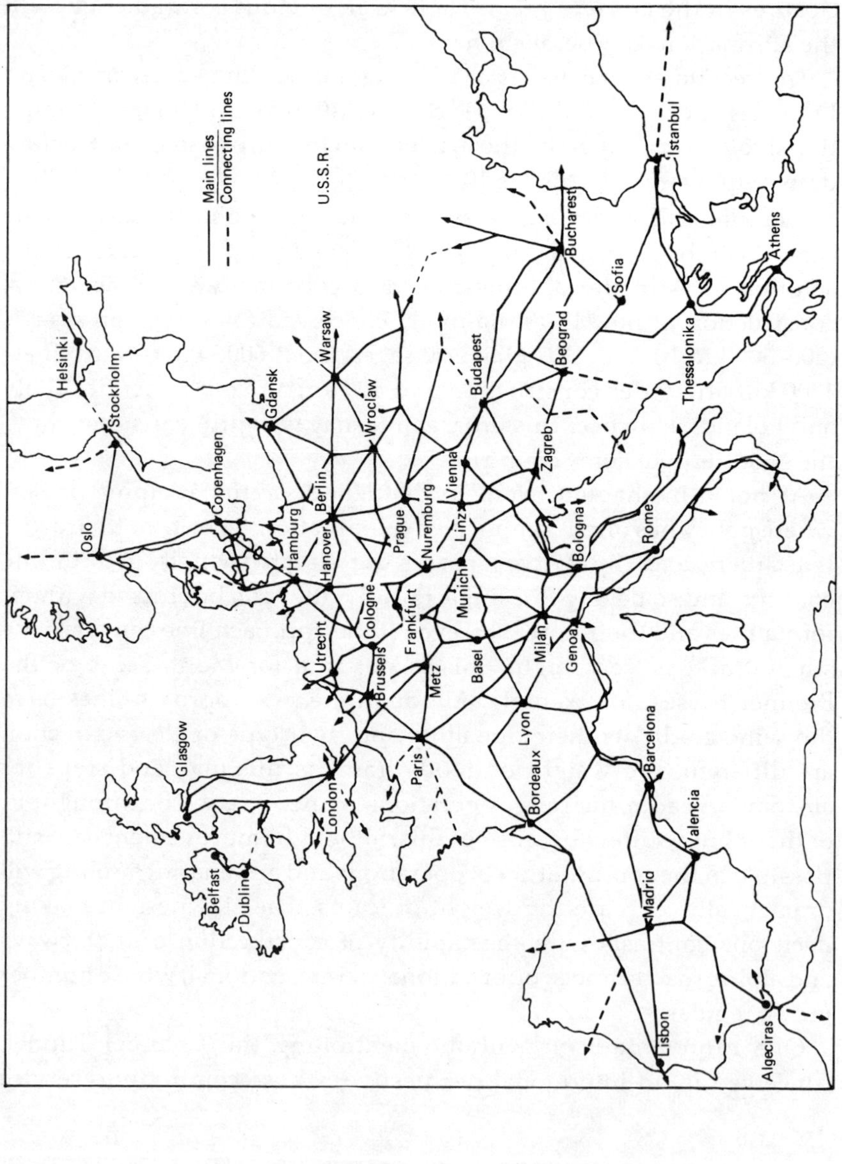

Fig. 10. Master plan for European infrastructure as drawn up by the UIC (international links only)

would transform land transport in one of the most populated and industrialised regions in the world. Linked to several motorways and high-speed railway lines, the Channel Tunnel would provide a new dimension to human exchanges and trade in the whole of Western Europe. This is one of the few construction projects that concern almost all of the member countries of the EEC. It is desirable that works which were deferred at the last minute in 1975 be finally undertaken as soon as possible. The development of the energy situation can only accentuate the intermodal role. In the same way, the projected tunnel just a few kilometres in length under the Oresund would make possible a continuous rail link to Scandinavia, and a bridge would do the same for Sicily.

European railways should not however be accused of megalomania. Their programme is very modest when compared with those of roads and aviation, but shows up very well because of the enormous potential capacity a railway presents. Thus rejuvenated, the European railway system should supply the community with an instrument suited to the demographic and economic structure of the next few decades and calling for intermodal cooperation based on an objective analysis of total costs and benefits. It will obviously be up to the railways to ensure the best use. At the international level, where traffic is by nature very diffuse, the future is linked to the development of communal sectorial action, and I have listed the first examples. Of course, such examples imply a basic consensus on the structure of costs as well as agreement to surrender sovereignty over tariffs; but convincing results recorded by the first subsidiary companies have opened the way.

Finally, it should be recalled that Europe, which saw the birth of the railway, has always been the centre of railway research and invention. All the changes which have made possible the renaissance of the railway in recent decades originated there: welding rails into long sections, electric traction using current at industrial frequencies, traffic running more than 300km/h, stability of the track–vehicle couple at high speeds, etc. The existence of a strong nucleus of researchers and a communal structure at the level of the macrosystem will ensure the continuation of this, but only retreat from chauvinism at the three levels of railwaymen, industry, and governments, will make it possible to optimise them.

Therefore the future of the railway in Europe only poses problems in the competitive sector of railways in countries with liberal economies, problems resulting from the absence of a coherent transport policy.

Considered objectively and in the light of new energy problems, this future cannot, in the interest of the community, be put in doubt, it being understood that the railway must be authorised to take final leave of the period of the monopoly, and from now on be limited to markets and routes where it offers the best method of supply, and perhaps the only one. The answer is up to governments and inter-governmental authorities.

64 The USSR

Although it comes under a single management, the Soviet system, the most important in the world, has the characteristics of a national macrosystem. In this archetype of the planned economy, the railway has an essential role in the transport system. In spite of the rapid development of other modes, rail is still by far the country's principal carrier. Altogether it carries 60 % of goods and 40 % of passengers. The execution of Five-Year Plans, especially in industry and agriculture, depends on the proper functioning of the railway, whose organisation and running methods are controlled by research into maximum efficiency.

641 *General characteristics*

The geography of the USSR lends itself particularly well to utilisation by rail: it is a massive continent of gentle relief, an extreme climate that renders river and road transport in some regions difficult and at times impracticable, and with human and natural resources distributed over truly vast distances.

The railway was first developed in the European part. Economic and strategic necessities soon caused it to extend into Asia, and the Trans-Siberian was completed in 1904. Since then the length of the system, which reached 70000 km in 1917, has regularly increased; it has doubled in 60 years, with most extensions in the east and south east of the country. The Second World War extensively disrupted the European part of the system, and when peace returned an intensive programme of reconstruction and gauge unification received priority. System extension programmes were pursued at a high rate, and construction in progress on the Baikal–Amur line, which will double the Trans-Siberian line over 3200 km of permafrost country several hundred kilometres to the north of the Chinese border, represents one of the greatest enterprises in the history of transport. The network grid is well developed in the European part, but spread out in Siberia; the major axes to Turkestan and Baikal have become corridors, while small lines are unknown.

642 *Technology*

Technology takes account of these general characteristics. The track is equipped with heavy rails, and a fundamental option has been adopted in limiting the maximum load per axle to 22 tonnes, which means that the track can be used by very dense passenger traffic under satisfactory conditions of maintenance and comfort. As a result, 3 or 4 axle bogies are fitted to a growing part of goods rolling stock, which besides being of very strong construction is relatively heavy. Steam traction has almost completely given way to dieselisation and electrification, and engines are built according to the same robust criteria. Standardisation, helped by the volume of stock (several million wagons) is very widespread.

643 *Operating methods*

These are based on the priority given to goods traffic and the maximum use of installations, and so involve very regular traffic. The maximum speed of passenger trains exceeds 120 km/h only in exceptional cases. Automation is widely developed (signalling, marshalling yards) and there is radio communication between locomotives, stations and traffic control centres on all the major lines. Trains follow each other at 10–15 minute intervals over long stretches of line; their frequency over the 9000 km of the Trans-Siberian is the same as that of some suburban lines outside peak hours on other railways.

644 *Goods traffic*

The volume (nearly 4000 million tonnes) and the average distance transported, about 900 km, are very favourable to the good use of equipment, all the more so because rigorous planning is also applied to loading and unloading operations. The turn-around time of rolling stock, about five days, is particularly low. The average gross weight of trains is more than 2700 tonnes, their length is limited by that of storage sidings – around 1000 m. Complete train loads running between private sidings make up most of the traffic; it seems that the gross load of the heaviest trains is not more than 10000 tonnes. Transport by container is also developing and at present reaches 40 million tonnes. Block trains with 52 container wagons on the land bridge between Europe and Japan by the Trans-Siberian run at a commercial speed of 25 km/h. International traffic represents only about 5 % of the total traffic. Its execution depends on external specialist commercial authorities who carry out the practice of dumping, should the occasion arise.

The goods efficiency of the Soviet network is thus impressive. With 24.5 million tonnes per kilometre per year of line it is seven times that of the US and ten times that of Europe. The absence of small lines plays an important role in this comparison. The average volume of kilometric traffic in the USSR exceeds that of most lines specialising in traffic of a single raw material (iron ore, coal) which only use very heavy trains and have no passenger traffic. The Soviet system thus transports more than half of the world's railway traffic on just 12 % of the length of the world network.

645 *Passenger traffic*

This is also very important. The intercity service consists essentially of long-distance trains (journeys of up to eight days) and night trains, composed of coaches with couchettes, either 'hard' or 'soft' class, and growing numbers of sleeping coaches. The infrequent day trains have a single class. The rolling stock is relatively heavy and the average weight per seat supplied may be two tonnes for a train of 1000 tonnes gross. Seat reservation is compulsory, which poses complex problems because of the possibility of re-utilisation on long journeys; automation of the system is in progress. Traffic density is very high on the main European links: up to 40 trains in each direction over long sections that are sometimes single-track (Black Sea Riviera), and even 20 trains on the Trans-Siberian as far as Lake Baikal. Commercial speeds are rarely more than 70 km/h. The fastest train between Moscow and Leningrad, a straight, flat route, runs at 93 km/h. A unit train capable of 200 km/h announced several years ago is not yet in service. In spite of the enormous development of aviation, the train remains an essential mode of intercity transport for Soviet passengers whose criteria for appraisal of time-value are specific, and traffic is increasing regularly. The absence of small lines partly explains why traffic exceeds the level reached in Europe per kilometre of line. Finally, suburban services are very frequent in big cities, and the largest cities have metropolitan subways.

646 *Organisation*

This conforms to the model described in chapter 6 for countries with planned economies. At the head is a Ministry of Communications, which is responsible for running the railway (the construction of new lines and the railway construction industry generally is the concern of specialist ministries) and has the same functions as a general administration and ensures links with Gosplan. Operation of the railway is divided into 26 regions of very unequal size; they are given great autonomy, and there is a third level of command.

The Ministry also has a very large Central Institute of Scientific Research, which directs the Experimental Centre at Tcherbinka. Professional training is highly developed and there are 12 railway institutes at university level. There are about two million railway workers.

International relations operate within the framework of the OSZD (see chapter 7).

647 *Prospects*

The financial situation of the Soviet railways is excellent, confirming the growing efficiency of the railway. Part of the surplus seems to be directly reserved for investment. The enduring problem of the network is to increase capacity, and in this it follows the general trends developed in the previous chapter, with electrification, automation, and concentration being the essential areas. Some double-track electric lines with automatic signalling are approaching saturation point; the network grid makes it possible to consider the idea of corridors. Elsewhere the possibility of constructing specialised high-speed lines so as to clear some existing lines for passenger traffic, is still being considered according to recent declarations. The Moscow–Kiev–Black Sea and the Moscow–Europe via Poland routes naturally come first to mind. So, in spite of distances, the rise of air transport and a certain development of the private car, Soviet planning continues to reserve first place for intercity passenger transport by railway. As things will stay the same for goods traffic, we can be sure of the future of the Soviet network.

65 **Asia and Australia**

Asia does not yet have a unified railway system. Its northern half is part of the macrosystem of the USSR just described, and its southern half is divided between three macrosystems which at present are not connected to each other: the Middle East, the Sub-continent and the Far East. I shall examine them in turn, as well as the Japanese and Australian macrosystems. There are also a few isolated railways (Burma, Indonesia, the Philippines, New Zealand) and a secondary Kampuchea–Thailand–Malaysia macrosystem, whose international activity is very limited. Although there is no unity among the railway systems one factor which characterises most of the continent is population density. The consequences of this are important for all the railways in the area.

651 *The Middle East*

The Middle East RMS includes the railways of the following countries: Turkey, Syria, Iraq, Iran, Lebanon, Jordan, Israel and Egypt. A metric gauge line Beirut–Damascus–Amman, recently extended to Aquaba, plays an insignificant international role. The standard gauge line in Saudi Arabia is for the moment isolated. The link between Istanbul and Teheran was finished only in 1971.

A geographical crossroads, the Middle East is part of the TMS. It has the advantage of a very long sea coast. With the exception of Syria and Iraq, physical geography has imposed severe line characteristics and two sea crossings by transport ferries, in Istanbul and on Lake Van.

The structure of the macrosystem still bears the mark of its origins, associated with Europe (the Berlin–Baghdad railway). It is a branched network; starting from Istanbul, it branches to Iran (and to the USSR), Iraq, and the whole of Syria, Lebanon, Israel and Egypt. There is no international grid yet, and only Turkey and Egypt have a national grid. A link from Iran to Iraq via Basra (about 50 km over flat country) was laid after the last war. Except for the Nile Delta, there are no secondary lines.

All the railways concerned, except that of Jordan, have standard gauge track and apply the norms of the UIC, of which they are members. The five railways at present physically linked to each other, Turkey, Iran, Iraq, Syria, Lebanon, have also formed, within the UIC, a limited Middle East group so as to improve the coordination of their activities. Operational problems are settled by the Taurus Conference and a pricing union called the CMO consolidates the railways, which adhere to the CIV and CIM legal conventions.

The internal traffic of the different railways is developing rapidly, while international traffic is very sensitive to economic and political events. In the goods sector the principal traffic flows, which have increased rapidly in the last few years, go to Europe and the USSR. The Turkish railway is the linch pin, with a transit distance of more than 2000 km, and the railway is in strong competition with water and road transport.

In the passenger sector, Egypt and Turkey have quite a full service (20 daily links between Cairo and Alexandria, 208 km apart). On the international level, a few through coaches run between Istanbul and Teheran, Aleppo and Baghdad, as well as Teheran and Moscow.

The outlook for this macrosystem is very favourable. The Middle East is one of the most typical examples of the vital role of the railway in a region where oil dominates the economy and which has a new road network of good quality and in constant development. In fact, it confirms the claim that a country cannot be industrialised without the railway. This is why the railway programmes of the Middle East are of impressive size. In Turkey and Egypt, which already have a basic grid network, plans aim to improve the capacity of the major routes, sometimes by constructing new sections of line, and by extending electrification, up to now limited to suburban lines. In the other countries, it is a matter of developing the railway on its present basis, as part of agricultural and industrial plans, and of starting to build a grid. Saudi Arabia is preparing for the construction of a system several thousands of kilometres long which will link it to the rest of the Middle East, and the future railway in Afghanistan should play an important national and international role. Finally, very rapid demographic growth – there should be a 70 % increase in population between now and the end of the century – makes the railway an essential form of passenger transport. One should not neglect to mention that tourism is already important (Nile Valley) and should see some spectacular developments (Egypt–Israel).

652 *The Sub-continent*

This macrosystem includes the systems of India, Pakistan and Bangladesh. The railway coverage of the Sub-continent is adequate because of a well-developed grid and the existence of very important corridors (Calcutta–Delhi–Lahore, Calcutta–Bombay, Delhi–Bombay). The principal lines are double track and electrification extends to the suburbs of big cities and to major links. Broad gauge (1.676 m) is most often used, though 45 % of the Indian railway and 70 % of that of Bangladesh are metric gauge. A recent technical and economic study has determined Indian policy on the subject: it has been judged desirable to modernise the metric track, and changes of gauge will not be carried out except to satisfy the needs of very specific interconnections. The macrosystem has the advantage of technical and operational standardisation developed for a century under the British administration.

The Indian railway, which is more than 50 000 km and employs about 1.5 million workers, is by far the most important and is in the front rank on a world level; it carries more than 90 % of the traffic of the whole macrosystem. It is itself divided into nine regional railways and the central administration stems directly (as in Pakistan) from a Ministry of Railways. The Indian railway has a centre of research studies and standardisation at Lucknow, with considerable resources.

The railway is the essential mode of transport in the Sub-continent, whose economy is partly planned in the sectors of population, agricultural food products and industry. India in particular has important heavy goods traffic (coal, minerals), moved by block trains. These networks face permanent problems of capacity, which make substantial improvement in the passenger service difficult. This nevertheless attains high quality (eight expresses daily between Delhi and Calcutta – 1445 km, one of which has a commercial speed of 87 km/h, seven between Lahore and Karachi – 1212 km, for example). Modern rolling stock has air conditioning and double decker coaches are being developed. Suburban transport is extremely important, notably in Calcutta, Bombay and Lahore. The first metropolitan subway is being built in Calcutta. Between 1970 and 1976, passenger traffic increased by 40 %; every new train is immediately filled.

International traffic, disrupted for a long time, was recently opened again between India and its two neighbours. It remains very small, although there are no technical problems and is unlikely to play a significant role except for the Pakistani network after completion of the link with Iran and the construction of the Afghan railway. The railways of the macrosystem are preparing for this by adapting their technical standards to those of the UIC, of which they are members.

The outlook for the railways of the Sub-continent is very positive. Industrialisation is developing quickly and distances are long, while heavy agricultural food transport plays an essential role. Finally, the population will be more than 1200 million by the end of the century; the gross national product, although expanding, is one of the lowest in the world and time-value is very low. The railway is the only mode of transport capable of facing up to needs of such size. The imperatives of capacity will continue to have priority, and they imply the construction, sooner or later, of new lines in the main corridors. A single example shows the scale of the problem. Over a distance equivalent to that from Brussels to Warsaw, the Calcutta–Delhi railway corridor directly serves around 100 million inhabitants; the aeroplane is only accessible to VIPs and intercity coaches are barely developed.

653 *The Far East*

This macrosystem includes the railways of the following countries: China, Mongolia, North Korea, South Korea and Vietnam.

The heart of it is the Chinese railway, for which recent information is now available. Its length, rapidly expanding towards the west, is about 5000 km, 1000 km of which is electrified. It properly serves only the east and north east of the country, where the grid is developing. Very severe geographical conditions make steep gradients necessary in central regions. The abundance of coal explains the predominance of steam traction. The maximum load per axle is 23 tonnes. Absolute priority is given to goods traffic, which approaches 1000 million tonnes annually, with coal at the head of products transported, mostly in block trains. Passenger traffic now reaches 800 million passengers, a low figure in comparison with the population, but there are no suburban services. The fastest link, from Beijing to Shenyang (840 km) is covered at a commercial speed of 84 km/h. Although it is constantly faced with problems of capacity, the Chinese railway nowadays occupies second place in the world and its expansion should be considerable.

The North Korean railway is around 4000 km long. The Mongolian network is reduced to the transit line from China to the USSR (around 1000 km). That of Hong Kong (KCR) is essentially suburban (35 km). The railway of Vietnam (around 1800 km) is under reconstruction. The railway of South Korea (KNR), nearly 4000 km in length, is intensively used and is in full expansion as much for goods as for passengers; it has modern equipment. There are metropolitan subways in Beijing, Seoul and Hong Kong.

International traffic is very low. Passenger trains run once or twice a week on journeys between Beijing and Ulan-Bator–Moscow, Beijing–Moscow via Manchuria, Beijing and Moscow–Pyong-Yang, Beijing–Hanoi, some of which are the longest in the world (9000 km from Moscow to Beijing via Manchuria, a journey of 6½ days). The through service from Guangzhou–Kowloon (Hong Kong) has restarted. International goods trains run between Mongolia and the USSR on the one hand, and between China, North Korea, Hanoi and Hong Kong on the other. The Far East RMS, which alone serves more than a third of the world's population and a region of enormous natural resources, is as yet imperfectly developed and exploited, but should see an exceptional expansion in the coming decades. The outlook seems unlimited and may well lead to the construction of specialised lines for high-speed passenger traffic, for as in the Sub-continent only the railway is capable of satisfying transport needs, which are as yet relatively low for a population which is likely to increase by 60 % in one generation to reach more than 1500 million people by the year 2000, at the same time as industrialisation is making rapid progress.

654 *Japan*

The Japanese RMS, has several special characteristics apart from its insularity. The national railway (JNR) is split into two independent systems: the original network with metric gauge track (1.067 m) and a high speed passenger network with standard gauge track, the Shinkansen. There are also a large number of secondary passenger railways and the main cities are served by highly-developed metropolitan subways.

Building the original railway through difficult country meant steep gradients and much engineering construction (bridges, tunnels and viaducts). A quarter of the railway is double track and a third is electrified. The choice of metric gauge was reasonable at the time because high speed was not important and investments could be appreciably reduced in a country where the value of space is particularly high. The extraordinary results obtained, especially for the suburbs of Tokyo, have now justified the choice and confirmed the potential of the metric track. But the development of traffic, linked to the rapid growth of population and of the gross national product, congestion of the big cities, and the cost and overcrowding of air transport made it necessary in 1955 to formulate a radical solution to increase the capacity of the Tokyo–Osaka corridor which, over a distance of 515 km, serves 40 million inhabitants. The great merit of the JNR is to show clearly the possibilities of a fast intercity system. An exceptionally high density of traffic made it possible to give the Tokaido traditional geometric characteristics, which though requiring a considerable number of tunnels, viaducts and bridges, proved profitable after only a few years. The Tokaido revealed the potential of the modern railway just as motorways revealed the potential of the road.

The essential characteristic of the Japanese RMS is the predominance of the passenger market. The railway carries 35 % of the national traffic, roads 50 % and air transport 5 %. The frequency of trains which are in block formation is very high on all lines. Suburban services are very well developed, particularly in Tokyo where the JNR transports more than ten million passengers daily out of a total of 25 million, including private railways and undergrounds. The metropolitan system of Tokyo is one of the most important in the world and has connections with some lines of the JNR and the subway. However the intransigence of ecologists has not yet allowed anyone to use the new airport at Narita in a satisfactory way.

The success of the Shinkansen is clear from a few figures. Inaugurated in 1964 between Tokyo and Osaka (515 km), extended in 1975 to Hakata (1069 km), it is now served by 275 daily block trains running between 6 a.m. and 12 midnight at a maximum speed of 210 km/h, giving a commercial speed of 163 km/h for the Hikari (Lightning) express. The total daily number of passengers was from the beginning counted in hundreds of thousands; it was more than a million on 5 May 1975 and reached the first thousand million after 12 years. This line alone transports ten times more passenger-kilometres than the 38 000 km run by Amtrak in the US. The experience of the Shinkansen was also very illuminating in evaluating the relationship between transport by air and by rail when the latter is by high-speed railway. According to Japan Air Lines, starting the service on the second section of the line has reduced the share held by air transport by half and its volume by more than a third. When the aggregate distance is the same, air traffic is disappearing almost totally, and at least one air service has been closed. So over 1069 km (7 hours by train as compared to $2\frac{1}{2}$ hours by plane) air traffic is being lowered by about 25 %. All this confirms the observations made in chapter 4. Finally, it is worth noting that the traffic on the Shinkansen alone brings in more than a third of the total receipts of the JNR and that its running efficiency is around 60 %. One of the fundamental factors in this result is the reduction in staff expenditure: 18 % against 55 % for the whole of the network. Modernisation is therefore paying off, especially when realised via a ladder effect.

Goods traffic on the Japanese RMS is not very important, for sea transport naturally holds first place with more than 50 %, followed by roads with around 35 %; the railway has to be satisfied with less than 15 %, and its traffic is constantly diminishing. Add ageing installations and equipment to the geographical handicaps and to the localisation of heavy industry on the coast and the outlook hardly seems favourable.

Japanese railway technology is among the best in the world. It has developed, apart from the Shinkansen, passenger rail motor trains for intercity block train services, pendular suspension systems (an electrified line in the mountain is entirely served by trains of this type), an electronic seat reservation network which is the largest and most sophisticated in the world, and has installed anti-noise devices. The JNR has a very modern research centre at Kunitachi.

In spite of brilliant results obtained, the economic situation of the JNR is worrying because of the uneconomic local lines: 40 % of the length of the network carries less than 5 % of the traffic. Closures are politically very difficult. Here again, the adaptation of the system's structures is indispensable. Also, the level of fares which until recently had to be approved by Parliament, was still very low; it was recently raised by more than a third, but the situation has not improved.

However, the government has justified its original decision in favour of the extension of the Shinkansen network, the only system capable of transporting people in a country which will have more than 130 million inhabitants by the year 2000. There are several thousand kilometres of high-speed lines planned for the north and west, and construction has already started. This programme includes a submarine tunnel 54 km long between the islands of Honshu and Hokkaido, which should be finished in 1984. It is a national Channel Tunnel. Thus the policy of passenger orientation of the JNR is being pursued, confirming the benefits of the express railway to the community. The construction of the Tokaido will remain a very important event in the technical and economic development of the railway.

655 *Australia*

A massive continent, nearly twice as large as Europe but 30 times less populated, Australia is a vast desert (except for the south eastern seaboard) but possessing huge natural riches. As a result, the density of railways is unequal. Most of the system is divided between two states, Victoria and New South Wales. Each state initially chose its own gauge which differs, with one exception, from that of its neighbours. There are three gauges: 1.60 m; 1.435 m; and 1.067 m, and in 1950 the Federal Government began the considerable undertaking of converting the whole of the long transcontinental route from Brisbane–Sydney–Melbourne–Perth to standard gauge tracks.

Technology is at a high level and in Australia there are mineral trains of more than 20 000 tonnes, electrified suburban services with double-decker coaches, car-sleeper trains and one of the most luxurious long-distance trains in the world, the Indian Pacific, which links Sydney and Perth (3961 km) at a commercial speed of 66 km/h.

The economic situation of the networks has worsened seriously, for modernisation has been very slow throughout the last decades and the railway structure in the south east is no longer adapted to the market.

The Federal Government, which directly administers some lines, has proposed that it should take over the railways of the different states, and has created an organisation for railway research and development. A global policy does not seem to have been defined for the near future, although some coordination is exercised by the Australian Railways Commission, created in 1975.

66 Africa

661 *General*

Africa does not have a unified railway system. The structure of its networks is still essentially linear or branched, starting from ports, with the exception of the Nile delta and a part of South Africa. This stems from Africa's economic and geographical characteristics, its immensity, huge deserts, and low population.

There are, however, two African RMS: the Maghreb, which as we have seen is part of the TMS, and that of Southern Africa, which will be described later on. The Egyptian railway is attached to the Middle Eastern RMS, which was examined above.

In Africa there are a few examples of international railway management: the Abidjan–Niger railway (Ivory Coast–Upper Volta) and the Tanzara (Tanzania–Zambia). Some others (Senegal–Mali, Kenya–Uganda–Tanzania, Ethiopia–Djibouti) have been subject to political vicissitudes. African railways have become aware of the convergence of numerous problems and have created the Union of African Railways (UAR) which will be described in the next chapter.

662 *The Maghreb*

This includes the three national railways of Morocco, Algeria and Tunisia. It has a branched structure, constituted by a standard gauge backbone of 2400 km linking Casablanca, Algiers and Tunis, and by the tributary lines serving ports and mining centres, sometimes up to the edge of the desert. There are practically no secondary lines. The main mining lines are electrified and rolling stock is of European type. The railways of the Maghreb are also members of the UIC and of the CIM and CIV conventions. A committee of railway transport in the Maghreb ensures their coordination.

Traffic is essentially national. The production of the three countries is effectively identical (essentially iron, phosphate, and citrus fruits). Passenger traffic is important. The operation of the 'Trans-Maghreb Express', Casablanca–Algiers–Tunis, is for the time being limited to the Algiers–Tunis section.

The railways of the Maghreb have considerable development programmes, linked to the industrialisation plans of the three countries. Several new lines must be constructed, in particular an international link from Tunis–Tripoli. The rate of demographic expansion necessitates the continued reinforcement of passenger services.

663 *Southern Africa*

This macrosystem consolidates the ten railways to the south of the River Zaire. Entirely of 1.067 m gauge, it is articulated around a backbone of 5100 km linking the Cape to Ilebo on the Kasai, a navigable tributary of the Zaire, which makes it possible to rejoin the ocean after another railway journey of 350 km between Kinshasa and Matadi[1]. Many branches link this backbone to the ports on the Indian and Atlantic oceans, the most recent being the Tanzara, which opened an outlet to Dar-es-Salaam (figure 11). The political situation has partially broken up this macrosystem and has considerably reduced

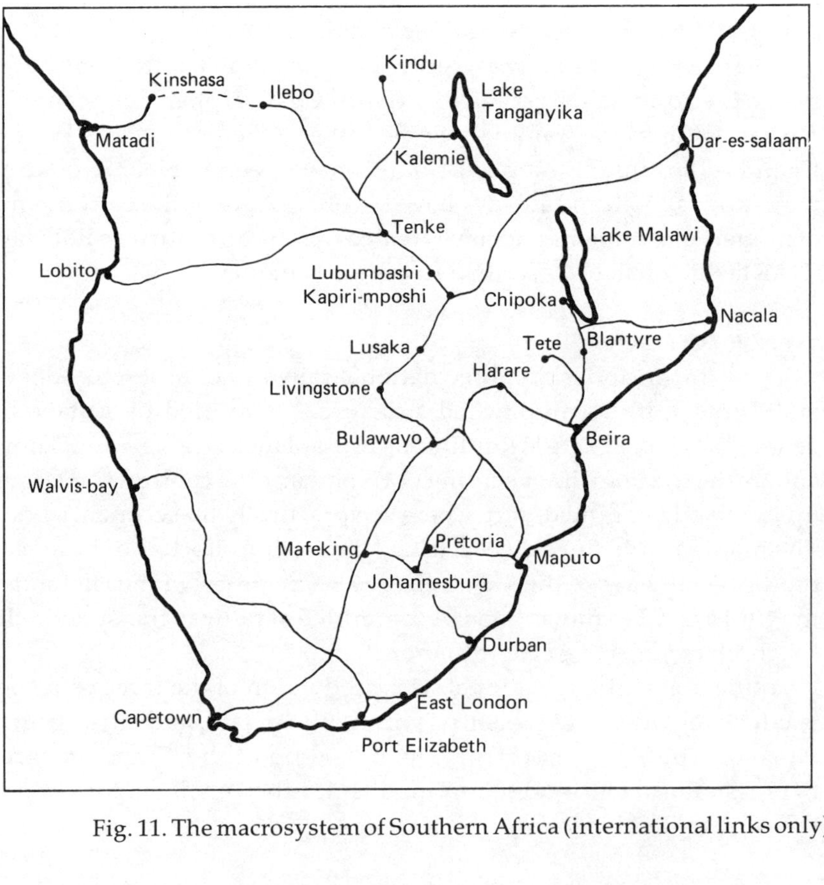

Fig. 11. The macrosystem of Southern Africa (international links only)

international traffic. No organ of coordination functions at present. The most important networks are members of the UIC, and most of them of the UAR.

The essential part of this macrosystem is the South African railway (SAR), whose technical standards are of common use. This railway is 22000 km long, with some grid development, and is partially electrified, through steam still plays an important role. It is the largest railway in Africa, and ranks technically and operationally among the most modern in the world. It provides the most convincing evidence of the feasibility of the metric gauge, as much for passengers (intercity links, suburbs, the luxury Blue Train (with running speed of 245 km/h in tests) as for goods (iron ore trains of more than 20000 tonnes gross on the Sishen–Saldanha Bay line, electrified by 50 kV single-phase current). The other networks are essentially linear and the majority of them are mostly used for mineral exports and goods transit. Passenger services are less important.

664 *Prospects*

The African railway system at present serves the principal centres of extraction of raw materials, the big towns and the main heavy industrial areas of the continent. In August 1978, the Ministers of Transport of the member states of the ECA adopted a very ambitious plan for the African railway, totalling 15600 km of new lines, mostly international. This plan aims to link all the existing railways and to end the isolation of the countries not yet served by the railway. It is doubtful if such a plan will be realised for a very long time because traffic exchanges are very low, the population small and the road network, parallel to most of the projected lines, may be gradually improved. Only a few mining lines, mainly national ones (Libya), seem to be justified in the short term. One might well presume that some of the present lines, isolated and short in length, with no heavy traffic and with good roads running in parallel to them, are doomed.

There are two priority problems that require international decisions, for they concern the future in a decisive way: the choice of gauge South of the Sahara and staff training. Africa south of the Sahara uses metric gauge track almost exclusively (1 m or 1.067 m) but changing bogies deals with problems of interconnections. It should continue to do so without hindering the development of transport potential in the slightest, although many wrong ideas, based on insufficient appraisal of the possibilities of metric gauge track, were suggested on this subject and some regrettable decisions were taken. A recent analysis, led by the group of the UIC operating railways with metric track has made it possible to determine the facts to do with this problem. Metric gauge track has the same potential as standard gauge track with the exception of very high speeds, the market for which is practically nil in Africa. The same loading gauge is used everywhere. The only criterion of choice is present and future compatibility, with the aim of developing macrosystems. This means that present gauges will be maintained and existing railways will apply the programmes for increasing capacity which were described in chapter 3.

Staff training poses an acute problem, particularly for the railways of countries which recently became independent. For the reasons described in chapter 3, it is desirable to organise this training in the country itself, though within an international framework, because of the low scale of individual needs. The UAR has rightly placed this matter among its priority objectives. They are currently preparing to open four training centres, which take language problems into account, at university level. It is to be wished that they receive the aid of all the interested organisations.

67 The Americas

The American continent does not have a unified railway system. Physical geography makes it possible to cover only part of the territory. Further, its immense size, the heterogeneity of its economic development and the political divisions of Central America have up to now obstructed any railway link between the two halves of the continent. Thus there are two isolated macrosystems.

The North American RMS includes the networks of Canada, the US and Mexico. The US and Canada, which have their own national RMS will first be examined separately. The RMS of South America includes most of the Latin American networks to the south of the equator. Between these two RMS there are a large number of isolated networks of little importance; they are very often linear and have metric gauge track. However Venezuela has just decided to construct a grid network with standard gauge track.

671 *The US: history*

This RMS is particularly interesting in several respects. It is a typical example of a national RMS, the predominance of goods traffic is absolute, and it is the last bastion of railway capitalism.

The development of the railway in a virgin continent normally produced systems constructed from ports with a linear structure to begin with, becoming branched later, and developing a grid only rarely (except in the north east). At the RMS level this grid was built only gradually. Such a policy has created a large number of corridors with competitive routes of the New York–Chicago type, and more especially in the Middle West, with a variety of gridding made up of long parallel lines and correctly called 'spaghetti' by John W. Barnum.

The nineteenth and twentieth centuries were the golden age of the railway, which truly 'made' the industrialisation of the US and contributed powerfully to its political unity. The big railway companies were the figureheads of capitalism and their history is dotted with financial battles and famous bankruptcies. There was no federal Ministry of Transport, but strict governmental control was exercised by the Interstate Commerce Commission (ICC), the legislation of the different states, and more recently by antitrust laws. The 1929 crisis sent many companies into bankruptcy and it was necessary to establish a bankruptcy statute to ensure that service be maintained. The railway networks of the US have thus never benefited from the total freedom of action generally associated with the most orthodox capitalism.

After a respite during the Second World War, in the course of which the railway played an essential role, the networks of the US were left physically undamaged. However, their equipment was outdated and they had to face totally free competition from the road, rivers, air, and pipelines. The north eastern railways, the densest and in closest competition with each other on the main corridors, rapidly saw their financial situation deteriorate. Permission to amalgamate was frequently denied under antitrust legislation, while those which did amalgamate were not always successful, as was shown in the resounding 'Wreck of Penn Central'[1]. The railways of the mid west and west were less dense and better off because of a certain amount of transference of industrialisation, and so generally did not have the same problems. The difficulties of the passenger service were particularly serious. Low population density together with the long distances between cities favoured the car and the aeroplane; on the other hand there were no small local services. The deficit on passenger traffic rapidly increased, renewal of rolling stock practically ceased from 1955, and trains were discontinued so that, except on the 'North East Corridor', Boston–New York–Washington, they were reduced to just one or two per day on a run-down network. The Pullman Company, which was responsible for most of the sleeping and restaurant services, disappeared. These brutal developments brought the structural heterogeneity of the US RMS into clear view, together with the need to formulate a federal transport policy.

The creation in 1966 of a Federal Ministry of Transports (DOT), together with the Federal Railroad Aministration (FRA), marked the beginning of a new era for the railway in the US. The first developments were the following:

(*a*) nationalisation of the intercity passenger service, under the National Railroad Passenger Corporation (Amtrak) and the formulation of a policy for fast passenger transport in the North East Corridor;
(*b*) amalgamation of the main networks of the north east by intermediary of the United States Railway Association (USRA), and the creation of the new Conrail network, which amounts to 'regional nationalisation';
(*c*) the development of research, with access to advanced foreign technology.

These different moves made it possible to formulate general legislation, Railroad Revitalization and Regulatory Reform Act (4 R act) of 5 February 1976. The 4 R act had the following objectives:

> To provide the means to rehabilitate and maintain the physical facilities, improve the operations and structure, and restore the financial stability of the railway system of the US, and to promote the revitalization of such railway system, so that this mode of transportation will remain viable in the private sector of the economy and will be able to provide energy-efficient, ecologically compatible transportation services with greater efficiency, effectiveness, and economy, through
>
> (1) ratemaking and regulatory reform;
>
> (2) the encouragement of efforts to restructure the system on a more economically justified basis, including planning authority in the Secretary of Transportation, an expedited procedure for determining whether merger and consolidation applications are in the public interest, and continuing reorganization authority;
>
> (3) financing mechanisms that will assure adequate rehabilitation and improvement of facilities and equipment, implementation of the final system plan, and implementation of the North East Corridor project;

(4) transitional continuation of service on light-density rail lines that are necessary to continued employment and community well-being throughout the US;

(5) auditing, accounting, reporting, and other requirements to protect Federal funds and to assure repayment of loans and financial responsibility; and

(6) necessary studies.

The new railway policy should, in particular, respect the following principles:

(*a*) reconcile the needs of the transporters, the consignors, and the public;

(*b*) develop competition between modes of transport to improve the quality of service and resume railway investment;

(*c*) increase the freedom of pricing of the networks, improve the flexibility of pricing structures, and reorganise railway tariffs bureaux;

(*d*) formulate directives to determine acceptable levels for railway receipts.

This law proposed truly revolutionary principles in respect of liberal philosophy by introducing ideas such as the public interest and planning. It has not been accepted without reservations in political circles as well as in the railway industry itself, and its implementation faces many obstacles. It is witness however to a new awareness of the function of transport, of the indispensable role of the railway in the economy of the US, and to the pragmatism of that country.

672 *The US: structure and operations*

After having reached 400 000 km the RMS of the US, now extends to 310 000 km only with geographical coverage varying according to the region for reasons given before. Almost all (98 %) the traffic is accounted for by 41 class 1 railways, whose annual income is more than $50 million. The law has up to now been opposed to the creation of transcontinental railways, whose merits are debatable[1]. The Conrail system, formed from the former Penn Central, is the largest with 32 000 km. Just 20 % of the RMS hauls nearly 70 % of the traffic, and 30 % of it only 1 %. Applying the 'Four R Act', the DOT devised a classification for the lines of class 1 railways in 1976 to initiate an outline plan for railway infrastructure. Four criteria were observed.

(*a*) traffic density (from 1.5 to 20 million tonnes gross annually);
(*b*) service to the main traffic centres;
(*c*) surplus capacity on some corridors;
(*d*) requirements of national defence;
using these criteria, six categories of line were defined.

This classification brings out certain characteristics of the RMS of the US. One is the insignificance of passenger traffic, a consequence of the human geography of the country. First category lines are defined as those which either haul a minimum of 22 million tonnes gross annually, or have links between the eleven principal traffic centres, whatever the tonnage; or finally those which run three express passenger-only services daily. Another is the importance of corridors. The plan identifies eleven of them, sometimes served by eight different lines, and up to 1000 km long. This clearly shows the infrastructure surplus.

Technical equipment reflects the physical and operational characteristics of the RMS: great climatic variation, low population density, predominance of goods traffic, high relative cost of manpower. Technology is oriented to robust construction and ease of maintenance rather than to performance. This leads to heavy rolling stock, greater standardisation – ensured by the AAR (see chapter 7) – and a limited number of series of vehicles. The fact that equipment is researched almost entirely by the builders has favoured these orientations. Diesel traction is the rule. Outside the North East Corridor and a few suburban lines, there are no longer any electrified lines; the mountain sections which were electrified during the age of steam have been dismantled in favour of diesel traction. However the recent development of the national energy policy may lead to a reversal of this tendency on some major lines. Goods rolling stock, fitted exclusively with bogies, is oriented to heavy traffic and the 100 tonne wagon. Automatic signalling is very widely developed, with long block sections.

Goods transport, which now represents practically the sole concern of the railways, benefits from favourable initial conditions: long distance of transport (an average of 900 km) and many markets for heavy raw materials (coal, minerals, etc.). Complete train load operations were developed later than in Europe, but then very rapidly. The present trend is to trains of 10000 tonnes (100 wagons carrying 100 tonnes each). Particular attention is being paid to automatic loading and unloading, carried out as far as possible on the train while it is moving (carousels). In parallel to this, some railways are experimenting with the low tonnage complete train load format. Unit loads are still very important and considerable investment has been devoted to modernising and automating big marshalling yards. Dispatch is slow however, for the high tonnage of trains limits their frequency and terminal operations in big industrial cities are very slow. Also the average use of wagons is low, 95 km per day in 1978, but has increased by more than 10 % in 10 years due to the multiplication of centralised traffic control systems. The average net tonnage per train, continually increasing, is more than 2000 tonnes. This is necessary largely because of regulations governing the composition of driving crew and accompanying staff. The unions have vigorously opposed, since the introduction of diesel traction, relaxation of the regulations in use at the time of steam traction (crews of five, journeys of 100 miles). Crew and staff still represent around 30 % of the manpower of the US networks as against less than 15 % on the large European railways. Recently, however, agreements have been concluded to reduce crew to four and in some cases three people, in return for some financial compensations. Combined transport, in the form of the container (COFC) and the road trailer (TOFC) is developing rapidly and easily because of the generous loading gauge. At present it represents around 15 % of the traffic. However, the role of the railway in goods transport in the US has declined continuously since 1920, except during the Second World War. Still more than 50 % in 1953, it was only 36 % in 1978. The development of pipelines partly falsifies this comparison, for the volume of railway traffic has increased by 26 % during this period. Recent years have been particularly good and the railways of the US beat their previous records in 1978 with 1310 thousand million tonne-kilometres transported.

Intercity passenger services run on only 38 000 km or 13 % of lines in use since 1971, and have been placed under the responsibility of Amtrak. Amtrak trains run on railway lines, but they provide their own traction, and because they operate the only passenger services, they pay the networks for track maintenance necessary to ensure acceptable comfort on lines where maximum axle loads of 30–35 tonnes are permitted. Trains usually run daily (except on some corridors), with journey lengths of more than 3000 km, though their average speed rarely reaches 80 km/h. Most of the monumental central stations have been greatly modified or sold. The North East Corridor, 734 km in length, is a special case. It serves more than 50 million people, its stations are well-situated at the centre of towns, and the line is mostly electrified. Also the passenger service has always been important.

Amtrak gave priority to its improvement and the 'Metroliners' at present make the New York–Washington journey (362 km) in $3\frac{1}{2}$ hours (a commercial speed of 103 km/h with four stops). The 'Four R Act' included precise directives for the development of the North East Corridor, and in 1976, Amtrak bought the track bed and track installations; it is undertaking a vast programme of repair and improvement financed by the Federal budget. Results are very encouraging, as the figures in table 6.3 taken from a DOT study, show.

Amtrak traffic in 1978 was about 6700 million passenger-kilometres (less than in Belgium or Switzerland) with an average distance of 350 km. This represents only 1 % of aggregate passenger traffic of the US, which mainly uses the private car (85 %) and the aeroplane (12 %). The railway and the coach, whose traffic is stagnating, share the remaining 3 %. Amtrak cannot achieve financial balance; its deficit, made up by Federal subsidies, was 1.68 times its receipts in 1978. The size of this figure led to a reduction in the density of the network at the end of 1979 (figure 12), although the oil crisis caused a considerable rise in intercity railway traffic for several months.

Car-sleeper trains were introduced very late in the US because the railways themselves were against the idea. The tenacity of the private company (Autotrain), which acquired its own rolling stock and constructed its own terminals, has made possible the inception of a daily service to Florida. Its traffic was 6 % of that of Amtrak in 1976 (average journey 1375 km). New routes are being studied. Finally, railways are being revived for suburban, airport and metropolitan services, which together represent traffic equal to that of Amtrak. In this way, metropolitan macrosystems have been developed, sometimes by local communities repurchasing lines from the major networks.

Table 6.3

		New York–Washington		New York–Boston	
		Rail	Air	Rail	Air
Development of the volume of traffic between 1969 and 1975		+32%	−33%	+87%	−39%
Distribution of traffic by mode	1969	25%	75%	11%	89%
	1975	39%	61%	26%	74%

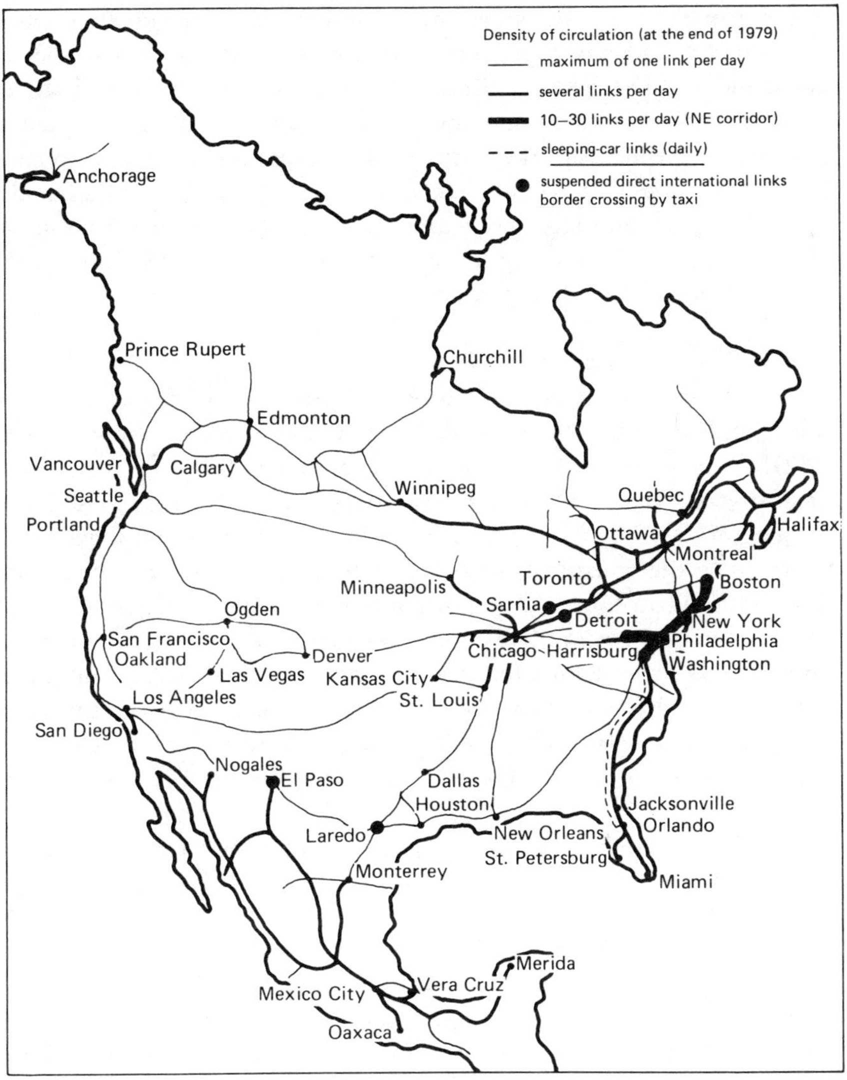

Fig. 12. The North American macrosystem (international links only)

Technical research, largely oriented towards safety problems and vehicle stability within the framework of the AAR, has taken on a new dimension with the creation in 1970 by the DOT of the Pueblo Research Centre. Conceived within the framework of a high-speed land transport project, this centre was to be devoted to research on new technology (linear motors, air cushions, magnetic lift, etc.), but technical and economic analyses showed that priority should be given to the conventional railway system. Most of the installations that were initially planned have not been constructed, and the centre is now devoting itself to research into railway dynamics. In particular, it conducts long-term testing of track. The relaunching of a high-quality passenger service on the North East Corridor showed up the technological lags in this neglected sector, and the Centre has called upon European and Japanese experience. For its new intercity rolling stock, Amtrak is extending the use of double-decker coaches which make full use of the loading gauge.

673 *The US: prospects*

There is a striking contrast between the increase in goods traffic and receipts, which in recent years have reached new records ($22 billion in 1978), and the decrease in net financial efficiency, declining in five years from 2.7 % to 1.6 %. This problem has a regional dimension: while the southern and western railways show spectacular progress in their results, the deficit of the eastern railways has more than doubled in four years and is more than the aggregate national profit. Two mid west railways have recently been declared bankrupt, and there is open debate between the policy of nationalisation, as in Conrail, or whether these railways should be taken over by more prosperous ones, thus eliminating 'spaghetti' routes. However, total investment in the railway industry has never been as high as it was in 1978.

Actually, it seems that there are plenty of reasons for optimism. The development of the energy situation is to the railway's advantage because it should increase coal transport. Potential competition from slurry pipelines is high, but the networks won their first fight in 1978. The cost of repairing the motorway network by 1996 has been estimated at $85 billion. The damage is largely caused by trucks, and the size of the figure gives pause for thought. Elsewhere the outlook for combined transport is favourable. Finally, a policy of 'deregulating' the railways, similar to that which created a stir among the airlines (at international level), was actively pursued by the Carter administration. However, the first applications, although limited, had a cool reception from clients, competitors and even railways themselves, whose point of view was summed up in November 1979 by William Dempsey, President of AAR:

> 'Many of the present problems of the railway spring from too much competition between railways, from overcapacity and

double use of installations. If the system is not restructured, it is counter productive to look for ways of developing inter-railway competition. Such measures serve only to redistribute the cake rather than to make it bigger'.

Adding to this the emphasis placed by the AAR on the need to balance the charges and subsidies of different modes of transport, it becomes apparent that the US railways have the same preoccupations as those in Europe (except for the bitterness of inter-railway rivalry characteristic of the national RMS) and their development in the next few years will be particularly interesting for the future of a 'liberal' railway policy.

674 *Canada*

The Canadian RMS is unique in that it consists largely of two railways covering almost all the country in parallel, one of which, the Canadian Pacific (CP) belongs to the private sector, while the other, the Canadian National (CN), belongs to the nationalised sector. The two lines sometimes shadow each other for long distances, thus creating long corridors. There are also some regional or mining railways. Only the south of this immense and largely uninhabited country is adequately covered, with about 70000 km of track. A few branches penetrate the far north, but a project for transporting oil by railway from Alaska to the US has not come to anything as yet.

The transport of goods is, as in the US, the main activity of the railways. The almost linear structure of each system facilitates operations using real time information centres, while their transcontinental character is a trump card as far as combined transport services are concerned. Technical equipment and methods of operation are similar to those in use in the US.

Passenger transport long ago ceased to be profitable for the same reasons as in the US. However, a service of satisfactory quality was maintained, due to a large federal subsidy, and re-organisation has just been decided upon by the government, according to a scheme inspired by Amtrak. Since 1 April 1978, the intercity railway services of Canada, with the exception of suburban services and some regional lines, are run by a new company called VIA Rail, which is a subsidiary of the CN. The purpose of VIA Rail is to ensure a satisfactory service by using the track of the CN or CP along the corridors according to the particular case (see figure 12). A programme for restructuring intercity links is being formulated. VIA Rail has benefited considerably from a programme of research conducted in recent years by the CN, notably into turbotrains and LRC trains ('light, rapid, comfortable'). The financial balance of VIA Rail is ensured by federal subsidies.

On the whole the situation of the two large Canadian railways is much healthier than that of their southern neighbours, and the extreme richness of the country in raw materials needing transport over long distances promises a favourable future.

675 *The North American macrosystem*

This includes the two preceding national RMS together with the Mexican system and the Alaskan line run by the DOT and linked to Canada by wagon-transporter ferries. The technological unity of the system is ensured by the standards of the AAR, of which all networks are members. Some Canadian lines belong to the networks of the US and vice versa.

The Mexican national railway (N de M) has been completely reorganised during recent years. It extends to about 15 000 km. A modernisation programme is being actively followed after a long period of neglect. Its prospects seem very favourable, not only for goods (the kilometric tonnage almost doubled between 1968 and 1978), but also for passengers, for the population of the country should double between now and the end of the century. The insufficiency of the metropolitan RMS of Mexico might well pose difficult problems however.

676 *South American macrosystem*

This RMS includes the railways of seven countries, Chile, Argentina, Uruguay, Paraguay, Brazil, Bolivia, and Peru, which together total 88 000 km of track. Geographical constraints are varied and severe. In Peru and Bolivia are the only lines in the world operating at altitudes higher than 4000 m. The escarpment between Santos and Sao Paulo for a long time required cable traction, and several lines use rack railways. Wagon-transporter ferries are used to cross some rivers and lakes. The equatorial forest is difficult to cross. Today the railway coverage of Brazil is far from complete while the region of Buenos Aires is somewhat saturated with lines inherited from former sometimes rival concessionary companies.

At present the RMS is something of a mosaic of railways with a few international links. There are several problems involved in its unification. In the first place is the unequalled diversity of gauges, of which there are four: 1.67 m, 1.60 m, 1.435 m, 1 m. Argentina has three, Chile

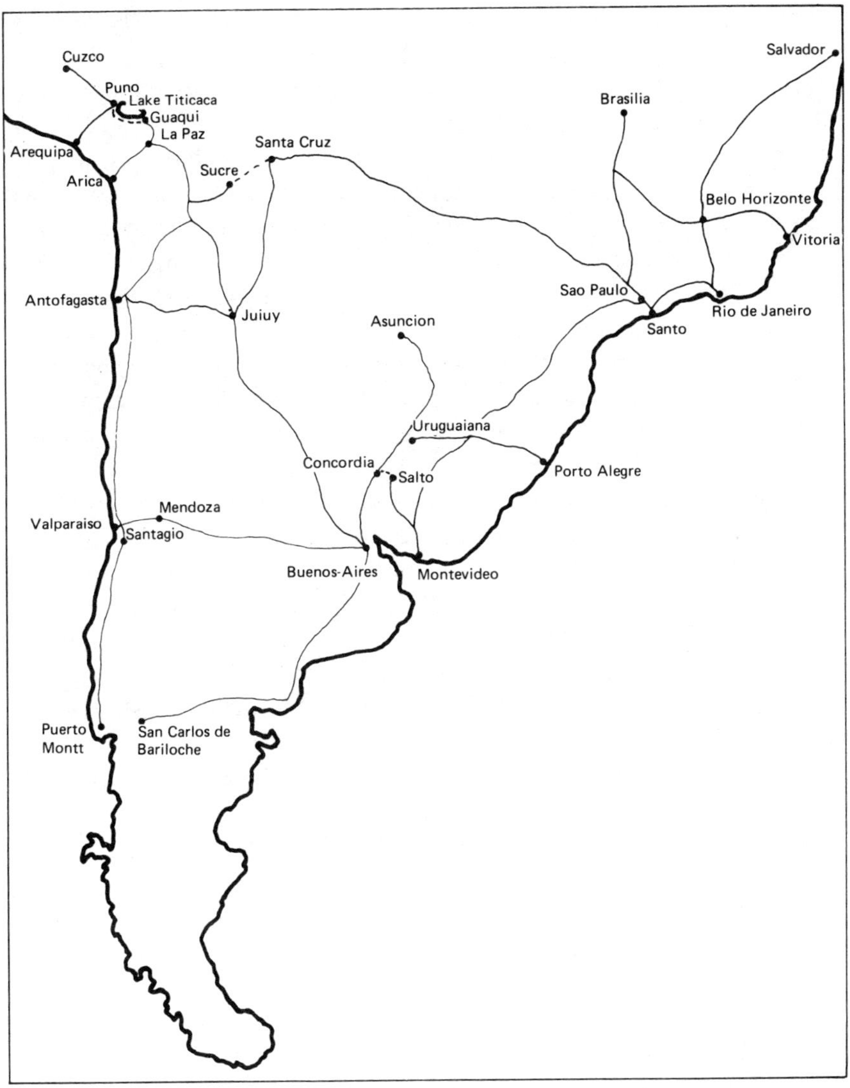

Fig. 13. The South American macrosystem (international links only)

and Brazil two, and there are no installations for changing bogies. The transcontinental link from Buenos Aires–Valparaiso requires two changes of gauge across the Andes. A more northerly link between Brazil, Bolivia and Chile does not require any, but involves a very long detour (figure 13). At present no technical unity exists. Outside the suburbs of Rio de Janeiro, Sao Paulo, Buenos Aires and Santiago, electrification of the main lines has been undertaken only in Brazil

(2300 km) and in Chile (900 km). Steam traction is still in service on some railways.

The two most important systems, Brazil and Argentina, have undertaken vast programmes of rationalisation (closing secondary lines) and development, based essentially on goods traffic. Heavy transport by complete train load is particularly important in Brazil. The metric gauge line Vitoria–Minas dispatches annually more than 60 million tonnes of iron ore in trains of 14 500 tonnes net, composed of 160 wagons of 90 tonnes gross, which illustrates the considerations mentioned in chapter 2 relating to gauge. Brazilian potential in heavy transport is considerable. Intercity passenger services are relatively frequent in South Brazil, Chile, in the east of Argentina, and in Uruguay. Elsewhere they are generally restricted to a slow daily service.

The only international transport of any importance travels on lines with metric gauge linking Bolivia to northern Chile, and Argentina and western Brazil. These represent only a negligible percentage of the total traffic of the RMS, but new links are either under construction (Argentina–Uruguay), or planned (Argentina–Paraguay, Bolivia–Peru) and these should develop traffic. The coordination of the activities of the railways of the South American RMS is carried out by ALAF (chapter 7).

The future of the railway in South America seems particularly favourable in Brazil, because of its enormous reserves of raw materials and of a demographic growth which will give by the end of the century a population of more than 200 million, the fifth largest in the world. It is possible that new high-speed lines offer the best solution for the Rio de Janeiro–Sao Paulo–Belo Horizonte triangle.

68 The main types of railway system

The tour of the world's railways just completed makes it possible to formulate a few general ideas on the present trends in railway development. One may recall that the railway originated with the block coal train, but that very soon after it was carrying passengers and almost all types of goods. Some railways have remained faithful to this original vocation, while others benefited from particular markets, or specialised completely. Out of this historical pattern have come several types of modern railway system, and I shall now give a general picture of their main characteristics.

The great majority of the railways of Europe, Asia, Africa and a part of Latin America have remained universal carriers. Subject to the development of markets, they continue to transport passengers and a variety of goods. However, their objectives are not all the same, and one can isolate two main groups.

The 'traditional' railways have as their primary objective the best utilisation of a supply which sometimes has difficulty in keeping up with demand. There is no room for isolated cases of high performance, which reduce the global capacity, and so little need for advanced technologies. The most illustrative examples are the USSR, India, China, and Eastern Europe, countries with planned economies, often of large size, and including areas of high population density where use of the private car is also limited. The railways of this group account for most of the world's traffic.

The 'sophisticated' railways, however, have as their primary objective to maximise quality of service to the user. They generally have sufficient margins of capacity to allow high performance in speed and they give some priority to the passenger service. The most illustrative examples are the four principal railways of Western Europe (West Germany, France, Great Britain, Italy); a certain minimum size is necessary for the transformation of the quality of service to become noticeable. These are railways of countries with liberal economies, which function in a fiercely competitive market, and they must be adapted to double sector operation.

The family of railways which give priority to goods traffic is illustrated by the US, Canada and Brazil. These railways serve vast, young countries, rich in raw materials, and with a population that is very unequally distributed. Since the last war they have had excellent road and air transport and they are all countries with liberal economies. Technology and operating methods are oriented to goods traffic; some passenger trains are tolerated, but their quality of service is mediocre if not poor. However some densely populated corridors do offer good quality passenger services. The ultimate stage of priority to goods is reached by specialised mineral lines.

The family of railways giving priority to passenger traffic is illustrated by Japan and the Netherlands, small countries with very high population density. Using advanced technology, these railways are expanding the metropolitan service format to intercity services of up to several hundred kilometres. Goods traffic runs only at night, but this is without commercial drawbacks in very small countries. The ultimate stage of priority to passengers is reached by metropolitan and suburban systems.

In the light of this classification, two fundamental parameters emerge, which articulate the activity of every mode of transport: geography and the economic system.

7

International organisations

Operating macrosystems requires, as we have seen, the support of international organisations. As a land-based system, the railway does not need organisation on a world scale as does air or sea transport. However, the advantages of standardisation and coordination are obvious. Since the last war we have seen the formation of political and economic links between countries. Transport is often among their objectives, resulting in the establishment of specialised bodies.

I shall now briefly review these world organisations – the United Nations at government level, the International Union of Railways at railway level, then the principal regional organisations, and finally the academic ones.

71 The United Nations

At the level of the General Secretariat, concern with transport is very limited. It comes within the framework of the Division of Resources and Transports and its principal activity is the financing of development projects, in particular railways, often in collaboration with the World Bank. By contrast, the activity of the UN in connection with railways is important at the level of regional economic commissions, whose headquarters are in Geneva for Europe, Beirut for Western Asia, Bangkok for South East Asia and the Pacific, Addis Ababa for Africa, and Santiago for Latin America. Specialist committees bring government experts together and study problems of common interest. They normally rely on the regional network organisations.

The International Union of Railways alone possesses the status of observer (category B) at the United Nations, and participates in this respect in these works, particularly within the framework of the regional commissions whose areas are part of the TMS.

72 The International Union of Railways (UIC)

721 *Objectives*

Created in 1922 in the wake of the League of Nations, the UIC had as its initial objective 'the unification and improvement of conditions for the establishment and operating of railways concerned with international traffic'. This objective was subsequently enlarged to include 'coordination and unification of action of international organisations adhering to the special agreement', and 'the foreign representation of the railways for the examination of common questions which concern them'. These adjuncts result in particular from the representative character of the UIC, recognised by the League of Nations from 1923 and confirmed by the United Nations in 1950.

722 *Admission*

Normal conditions of entry are the following: 'have at least 1000km of lines, with standard or broad gauge track, situated in Europe, or having rail links with the lines of the UIC, and serving public traffic in passengers and goods'. Some railways which do not fulfil all these conditions, such as urban or suburban railways or even transport enterprises which are not railways, may also join under certain conditions. In practice there are two categories of member railways:

(*a*) the nucleus railways, which make up a continuous geographical whole. This nucleus was the TMS when the UIC was founded. The Soviet railway left the UIC in 1947 and the railways of Northern China in 1948; on the other hand, all the railways in the Middle East have gradually been connected with the TMS, similarly those of the Subcontinent, and have joined the UIC. The total number of railways belonging to the nucleus in 1979 was 39;

(*b*) remote railways, essentially concerned with information, research and standardisation. Their number, which increases regularly, is at present 25 and they are distributed over all continents. In particular, the railways of the People's Republic of China returned to the UIC in 1979 as a remote railway.

723 *Organisation*

The UIC is governed by a board of management comprising the 21 general directors of the nucleus railways. The most important ones are permanent members, the others take part via a system of rotation. The president is elected every two years from among the board of management. The preparation of decisions, coordination and foreign relations are the task of a full-time secretary general, assisted by deputies from the member railways. The headquarters of the UIC is in Paris.

The fundamental objective of unification is periodically brought up to date through the choice of concrete priority objectives, to be realised through a programme of action which sets out all research and its terms of reference. The execution of studies is allocated to committees and a research and experiment office for which groups of experts provided by the member railways work part-time. The results are presented in the form of 'leaflets' giving recommendations or information or obligations. Together they constitute the UIC Code, which is being used more and more as a reference for all railway projects in progress throughout the world.

724 *Restricted application programmes*

It has quickly become apparent that even among the nucleus railways some problems were of a limited nature and that it would be easier to deal with them within a limited framework, while keeping to the principles and in coordination with the general policy of the UIC. After the Second World War, this approach developed, particularly after the creation of international bodies at government level. These were concerned with the organisation of transport within the various geographical regions. Apart from this, interest in specific sectorial action has rapidly become apparent. This is possible due to the flexibility of the statutes of the UIC, which allow for programmes of 'restricted application', and which cover a very varied field:

(*a*) pools and groups (Euro-wagons, sleeping-coaches, pallets, Trans-Europ-Express, etc.);
(*b*) subsidiary companies (Interfrigo, Intercontainer);
(*c*) economic/geographic railway groups (the railways of member countries of the EEC, the Middle Eastern railways, tropical railways with metric track, etc.).

The geographical extension of the nucleus of the UIC naturally leads to the development of restricted application programmes.

725 *International coordination*

This is excercised in many regions, starting with Europe, where some organisations established prior to the UIC have been maintained up to the present, either at government or at railway level or both. Justification for the autonomy of these bodies which date from a time when the majority of railways were private and intergovernmental bodies did not even exist, today appears to be absent. At the end of 1978, the UIC decided on the integration of the RIC and RIV unions, created before the UIC and charged with the problem of goods and passenger rolling stock exchange.

The UIC also maintains an official link with the International Standards Organisation (ISO) and the International Telephonic and Telegraphic Consultative Committee, which functions within the framework of the International Union of Telecommunications.

It also works in association with numerous professional bodies such as the International Chamber of Commerce (ICC), the International Federation of Transit associations (FIATA); the world Federation of Travel Agents Associations (FUAAV), the International Union of Associations of Private Wagon Owners (UIP), and the International Road Union (IRU).

Finally, the UIC cooperates, in various forms, with regional organisations which are going to be examined now.

73 Regional intergovernmental authorities

Only those which play an active role in the railway field will be mentioned here.

731 *The European Conference of Ministers of Transport (ECMT)*

Founded in 1953, and partially taking over from the organisations created in 1945 to put European transport back into operation again after the Second World War, it has the following objectives:

> 'To take all measures designed to realise, in a general or regional framework, the best use and the most rational development of internal European transport of international importance;
>
> 'To coordinate and promote the work of international organisations concerned with internal European transport, having taken into account the activity of supranational authorities in this field.'

The ECMT at present includes the 18 European countries with liberal economies, as well as Yugoslavia. It has no authority in the fields of sea and air transport and, because it does not have powers to compel, acts through resolutions.

Having taken into account the degree of cooperation reached by the railways of member countries of the UIC, the ECMT did not need to promote new agreements to facilitate international traffic. On the other hand, this was the idea behind the founding in 1955 of Eurofima, a financing company for railway equipment standardised by the UIC, which promotes leasing schemes. Eurofima consists of 16 railways and has very rapidly managed to acquire a high level of confidence in the international capital market, which makes it possible to carry out its operations at very reasonable rates.

The ECMT played an important role in spreading the principle of the standardisation of railway accounts, the definition of obligations of public service, and awareness of the delay in railway investment as compared with those in other modes of transport. It is closely concerned, in conjunction with the UIC, with all the main questions of the modern railway (automatic coupling, etc.) and from time to time organises conferences. In 1979, the Parliamentary Assembly of the Council of Europe 'pledged the ECMT to give priority to the development of transport by railway and waterway, which use less energy than the other modes of transport and, in particular, to encourage combined rail–road transport'. This confirms the new spirit of thought in liberal Europe.

732 *The European Economic Commission (EEC)*

The EEC was instituted by the Treaty of Rome of 25 March 1957; it now includes ten countries of western and central Europe[1], forming a geographical block. Transit through two non-member countries, Switzerland and Austria, is however essential for north–south traffic.

The mission of the EEC is

> 'by the establishment of a common market and by the gradual bringing together of the economic policies of member countries, to promote harmonious development of economic activities in the whole of the community, continuous and balanced expansion, increased stability, accelerated raising of living standards and closer relations between the countries it unites'.

Among its objectives is to 'draw up a common policy in the field of transport', notably by establishing common rules applicable to international transport and by suppressing discrimination. The provisions of the Treaty of Rome apply only to land-based transport, with the very important exception of navigation on the Rhine, which is still ruled by the Act of Mannheim of 1815; air and sea transport are also excluded, but the Commission is thinking of the possibility of incorporating them too.

Proceedings were directed after 1961 by a Memorandum of liberal philosophy, which aimed to progress on two fronts in parallel:

(*a*) harmonising conditions of competition between enterprises and modes of transport;

(*b*) freeing the transport market from institutional or quantitative restrictions on traffic within the EEC.

These directives were applied effectively for about ten years and allowed great progress to be made in the fields of fares, the harmonisation of certain important social distortions and above all, the standardisation of accounts (where several important prescriptions are still optional). However, the liberalisation of the market, which was of special concern to road transport, was relatively modest (Community quotas). But since 1972 there has been a marked lack of progress, due to various causes such as the enlarging of the EEC from six to ten members, with a resultant re-emergence of national rivalries, and an economic and monetary crisis. In spite of a Memorandum from the advisory commission in 1973, recommending the promotion of harmonisation so as to allow a certain degree of liberalisation, and concentration on the tariffication of infrastructures, the steps taken have been very slow, which has resulted in a worsening of the situation of railway transport as compared to road transport.

Some action in keeping with the spirit of 1961 has been taken, but has not yet become effective, while the liberalisation of road transport has progressed very quickly. However, some signs offer hope of redressing this balance, which seriously affects competition. A few EEC transport ministers are beginning to sound the alarm bell and call for a genuine common transport policy. At the beginning of 1979, the European Parliament in its turn came out strongly against the delay in devising transport policy and specified a certain number of priority actions, notably in the fields of harmonisation, prices and capacity. A Memorandum from the Commission in 1979 stressed the necessity for a Community policy for transport infrastructures, in particular for financing. This document largely rehearses, for the railways concerned, the discussion of motivations and priorities given above in chapter 6. Some hope for action in the short term is offered by a body of the EEC, the European Investment Bank, which has financed some railway projects in the past, at national and community level.

Thus the experience of the EEC confirms at the international level the difficulties encountered at the national level by the member countries trying to define a coherent liberal transport policy.

733 *The Council of Mutual Economic Assistance (CMEA)*

Created in 1949 within the framework of countries with socialist economies, the CMEA has for its objective the development of cooperation and planning in all areas of the economy. At present it includes the USSR, the six European socialist republics, Mongolia, Vietnam and Cuba. Its headquarters are in Moscow.

In 1971 it adopted a 'programme for the deepening improvement of cooperation and for the development of socialist economic integration' of its member countries.

Among its permanent bodies is a conference of transport and navigation organisations. For the railways, cooperation is effected by the OSZD.

The CMEA has recently prepared an integrated plan for international transport using combined techniques and coordinated programmes for modernising international axes. It has also developed some apportionment of tasks in the construction of equipment.

734 *The Central Office for International Railway Transport (OCTI)*

Created in 1890, the OCTI is concerned with legal objectives. It formulates international sovereign conventions in respect of responsibility for international railway transport: the International Convention on Transport of Passengers and Luggage by the Railway (CIV), and the International Convention on Transport of Goods by the Railway (CIM). Since the creation of the OSZD[1] these agreements are no longer applicable to transport involving only the member railways of that body. The member countries are those whose railways are part of the UIC in Europe, the Middle East and in the Maghreb (Morocco, Algeria, and Tunisia), and liaison with the railways is ensured by the International Committee on Railway Transport (CIT), which works closely with the UIC. The head office of OCTI is in Berne.

A clause of the CIM, which dates from the time of the monopoly and private companies, gives the forwarding agent the right to choose an itinerary, which does not necessarily correspond to the most rapid or most economic routing while international road transport is completely free in its itineraries, giving it much more flexibility and efficiency. This results in a distortion of the efficiency of international traffic, and the question seems to merit a new approach.

74 Professional regional organisations

741 *The Organisation for the Collaboration of Railways (OSZD)*

Created in 1957 by the socialist countries the OSZD, as noted in chapter 5, combines governments and railways. At present it has 13 members: the USSR, the six socialist republics in Europe (which are also members of the UIC), Albania, Mongolia, China, North Korea, Vietnam and Cuba. Its head office is in Warsaw. Its objectives and methods of working are practically the same as the UIC, but it has incorporated all the activities which, for the UIC, are still the concern of separate organisations. In the legal domain, the OSZD directly administers the two conventions of SMGS and SMPS, as well as the unified tariffs of transit (MTT) that apply to rail links between its members. The OSZD is also active in road transport.

The directing authority is the Conference of Ministers, and the executive authority is the Committee, with a representative from each member railway. The committees of the OSZD are presided over by full-time railway officials, and they publish leaflets similar to those of the UIC. Groups common to the UIC/OSZD have been created to study problems posed at the level of the TMS, notably those involving engineering and computing.

742 *The Association of American Railways (AAR)*

Founded in 1934, the AAR amalgamated a large number of specialised railway organisations in the US. Its essential objective is to coordinate the efforts of its members in all fields relating to the national transport policy and to act as an official representative of the industry to government legislatures and legal authorities.

The AAR is open to 'Class I' and 'Class II' railways (whose revenue is less than $50 million) as well as the local networks of the US. Its international nature results from the inclusion of the Canadian and Mexican railways. The AAR is directed by a council elected from among its member railways, and executive affairs are administered by a full-time president. Its headquarters are in Washington, DC.

The activities of the AAR are divided into departments, and there is a research centre in Chicago. The department of Operations plays a very important role in data processing, administering the telecommunications network TRAIN, which deals with exchange of wagons between railways and is the first inter-railway data bank in the world. The AAR works in close liaison with the Interstate Commerce Commission (ICC) and the Federal Department of Transportation (DOT).

743 *The Association of Latin American Railways (ALAF)*

Founded in 1964 by the Latin American railways the ALAF aims to develop cooperation and coordination, to promote technical and operational standardisation and thus the integration of the railway industry, and to research all methods of facilitating traffic.

There are both active members, divided into two classes according to whether the networks are linked to each other (the South American RMS) or not, and associate members (metropolitan lines, industries, various agencies). The ALAF comprises the railways of 11 countries, with headquarters in Buenos Aires. The administering body is a consultative council, which consists of representatives of all the active members, as well as a representative of the associate members. The executive body is the Secretariat. Studies are carried out by specialised technical and regional groups.

744 *Union of African Railways (UAR)*

The UAR was founded in 1973 within the framework of the United Nations Economic Commission for Africa and the Organisation of African Unity (OAU). It 'seeks to ensure the unification, development, coordination and improvement of the railway services of its members by connecting their railways to one another and to all other means of transport linking Africa with the rest of the world'. It seeks further to establish useful relations with other interested institutions.

Its members must belong to countries that are part of the OAU, a political organisation; it does not cover the whole of the continent. At present the UAR comprises 28 railways, which are mostly isolated, and is divided into four subregions. The directing body is a council, which includes a representative from each subregion, while the executive body is the general secretariat. The headquarters of the UAR are in Kinshasa. Work is executed through seven commissions.

The initial programme of the UAR includes a study of the main links needed to end the present isolation of the railways, the standardisation of equipment and training of executives.

745 *Other organisations*

An Arab Union of Railways was recently founded in 1979; it aims to include the railways of member countries of the Arab League. Its headquarters are in Aleppo and, like the UAR, it is a political agency. An Asian Union of Railways is being created under the agency of the United Nations Economic Commission for South East Asia and the Pacific.

Geopolitical divisions have led to the founding of the numerous regional organisations described in this chapter, but they do not always correspond to the structure of the macrosystems which are the foundation of international traffic. This is particularly so for the TMS, as was shown in chapter 6, and confirms the importance of the unifying role and intercontinental coordination of the UIC.

75 Railway Congresses

751 *The International Railway Congress Association (IRCA)*

Founded in 1885, the IRCA aims to encourage the development of railway transport by intensifying exchanges of experience between its members through all necessary means, notably:

(*a*) by holding regular congresses and other meetings with more restricted participation, either of a general or a specialised nature;

(*b*) by communicating information on particular problems to its members;

(*c*) by publishing technical reviews.

These activities are conducted in closest possible cooperation with other international railway organisations.

The IRCA is open to railways more than 100 km in length, to countries and to various railway authorities, and now includes 85 railways, 30 governments and 16 organisations. It is directed by a managing committee, with headquarters in Brussels, and direction is Belgian.

The essential activity of the IRCA is the organisation, in principle every four years, of a congress to review the situation in the different fields of railway activity. The congresses have an informative capacity. Because most members of the IRCA are also members of the UIC, it was decided in 1968 that the congress would henceforth be organised jointly by the IRCA and the UIC. The same goes for the publication of the review *Rail International – Schienen der Welt*, which has become the official communal organ, and of the *Summary of International Railway Documentation*.

752 *The Pan American Railway Congress Association (PARCA)*

Founded in 1906, the PARCA aims to promote the development and progress of railways of the American continent through the organisation of regular meetings and the collection and diffusion of documentation.

The PARCA is open to railways and governments situated in America as well as to different enterprises linked to the railways. At present it consists of 21 countries, with headquarters in Buenos Aires. It is headed by a permanent commission and includes national commissions. The essential activity of the PARCA is the organisation, in principle every three years, of a congress on the same formula as that of the IRCA, as well as the publication of a bulletin.

753 *The International Union of Public Transport (UITP)*

Founded in 1885, the UITP aims to study all the problems relating to public transport industries (buses, trolleybuses, tramways, metropolitan subways of local and regional concern). Its headquarters are in Brussels.

The UITP is open to private enterprises and public transport authorities of the whole world, as well as associate members. It therefore includes a number of railways notably in the Committee of Metropolitan Lines. Its activities include the administration of a documentation office, the publication of a review and the holding every two years of a congress which formulates its recommendations.

POSTSCRIPT

In this analysis, limited to the essential features of the present state of the railway, I hope to have convinced the reader of the reality of the renaissance of the railway mentioned in the introduction. On a world scale, it is an indisputable fact that the railway is doing well, a fact that is all the more remarkable because there has never been in the whole history of transport, such an abundance of continental modes of transport, land-based or airborne, discovered or revived in less than a century. Of course, in Western Europe, the Middle East, the US and Japan there are areas suffering from what could be called 'slow-developing post-monopolyitis.' The symptoms of this affliction are distinctive. The sick railway belongs to a highly-industrialised liberal economy, its body seems worn out but it still has all its faculties, especially its creative ones, leading to chronic problems of balance. Surgical treatment exists, and is desirable for many of the sick railways. But it is legally impossible for them, so the patients must go back to their guardians, who hesitate to do anything in the face of strong reactions from an ill-informed public and who therefore often prefer annual injections or tonics. A precarious balance is thus maintained, but the morale of the sick railways declines and the taxpayer revolts at the growing cost of the sickness benefits. It is to be wished that some of these areas where the sickness is endemic (fortunately it is not contagious) be cut back more vigorously. In any case a tree cannot hide the forest.

Economists may well draw constructive conclusions for the future from the difficulties experienced by a public service trying to get rid of the after-effects of a century-long monopoly. The size of these difficulties is itself evidence of the great importance the community attaches to the railway. Indeed, on the eve of the twenty-first century, as world transport demand grows under the pressure of increasing population and rising living standards, the volume of transport cannot but increase while available space shrinks, some energy resources diminish, and the society demands more clean air. The distinctive characteristics of the railway provide a ready answer to these problems: guidance and the convoy system increase the potential capacity of the ground area, electric traction can be served all primary sources of energy and does not pollute, cybernetics increase output and safety, growing efficiency reduces costs as traffic increases. The tendency of the other modes of transport to approach the conditions of physical guidance, the plans for non-conventional transport systems, the diversification of the use of pipelines – the only other guided mode – show that, if the railway did not exist, it would be imperative to invent it.

Is it reasonable to go further in a medium term forecast? The answer would normally come from economists, for transport is only one of the instruments of the economy, and practically all countries have a place for it in their plans. While waiting for their response, I shall only mention the two developments least subject to dispute, demography and energy costs. As far as demography is concerned, I noted in chapter 6 that, with the exception of Europe, the following decades will see a rapid increase in population, especially in regions where the gross national product does not, in the short term, show room for significant development of individual transport. The future of the passenger market for the railway is therefore assured, suggesting considerable investment in metropolitan and suburban lines, not to mention high speed lines. In spite of their cost, these investments are indisputably the most advantageous for the community. Another consequence of demographic growth is an increase in the volume of transport of agricultural food products, for which the complete train load offers the most economical mode of surface transport. There are many reasons for the increase in the cost of energy, and it is no longer doubted. I mentioned before the privileged position of the railway in this respect. Part of intercity traffic in countries with liberal economies should thus rediscover the railway. One should note however that a decline in the size of the car industry in countries where saturation of the market has almost been reached would in its turn have negative repercussions on railway traffic, because the car industry has become one of the railway's most important customers. In countries with planned economies, which already haul most of the world's railway goods traffic, the railway will continue to progress, even at the cost of heavy investment in capacity.

Finally, one can be ensured of the development, on a world scale, of the market for raw materials so vital to the needs of a population increasing in number and income. Albeit that the movement by land or sea of billions of tonnes of bulk ores and crude liquids, over hundreds or thousands of miles at speeds of 40 or 50 kilometres an hour, does not constitute one of the most exalted activities of an age in which information crosses space at the speed of light. But such transport is still, in the present state of industrial technology, indispensable to the community. One recalls Aldous Huxley's words: 'Moving bits of stuff from one end of the planet to the other, that is man's occupation'. This remark still contains a great deal of truth.

The problem for tomorrow's railway is not then one of survival, but of optimising its role as a complement to the modern version of the other modes, at the lowest aggregate cost to the community. The captive market sectors are already known: metropolitan subways, mining lines, oil pipelines, great natural waterways, small feeder roads; the optimisation is to be sought at the level of the multimodal intercity network. It depends in the first place on coordination of investments within the framework of a coherent global transport policy. 'Give us a good transport policy and I will give you good railways' to paraphrase Baron Louis, could be the maxim of many a railway manager, and highly-motivated railwaymen, faithful to their long tradition of service, have this at heart in their efforts to do the utmost for the community. It is also comforting to observe the revival of interest among young people in jobs on the railway, and it is specially to them that I direct these words from a posthumous work of Louis Armand[1].

> The railway is a vast crossroads where not only the fundamental disciplines, such as physics, chemistry, mathematics, and economics, meet but also others to do with actually operating the train – sociology, psychology and the like. I shall have achieved one of my aims if I have raised a few ideas which stir in others the desire to become, in the field of railways, 'humanistic engineers'. These, far from being just engineers (whose prerogatives I respect) have the duty to bring all these disciplines together, so that the gradual disappearance in mechanical creations of all anthropomorphism will not weigh upon us as some cold necessity, and so that the common man does not feel exiled among things which though still obedient to him pass his understanding as far as how they work. They will intervene to ensure that technology does not serve as one

of the instruments of power and they will thus counterbalance the activity of technocrats, a term in which, for sharp ears, the second syllable rings surprisingly stronger than the first. They will strive so that man, a creature worthy of respect, retains the respect of his own creation.

NOTES

Intro. A survey made in 1971 by the European Conference of Ministers of Transport is particularly enlightening on the distribution of transport infrastructure funds; and it highlights the growing backwardness of the railway in this field.

Intro. Joel de Rosnay, *Le Macroscope* Paris, Seuil, 1975.

111 The pneumatic metropolitan train, a birail with lateral guidance, needed an auxiliary rail–wheel guide to enable it to change track. The VAL system (Automatic Light Vehicle), whose first line is under construction at Lille, makes use of a segment of the central rail making a groove and a double tread for points.

112 Rail–tyre adhesion saw some development around 1935, especially in France (the 'michelines'), but was soon abandoned because it allowed only very low axle loads and required a large tractive effort. It was recently taken up again successfully on metropolitan lines, using longitudinal sleepers much wider than rails.

221 The cumulative length of railway bridges is only 1.3 % of its total length as against 13 % on the Direttissima from Rome–Florence and 33 % on the Shinkansen from Tokyo–Osaka.

222 The construction of a very long tunnel in a mountainous region raises a delicate ecological problem: where to put roughly four million cubic metres of debris from a double-track tunnel 50 km in length without altering the site or the drainage?

323 An exhibition organised by the Centre Pompidou in Paris at the end of 1978 on the theme 'The age of the station' illustrated this evolution of the sociological role of the station in a city.

363 In the US where the display of income is one of the accepted features of liberal democracy, the era of the railway tycoons in the nineteenth century has gone. However, the salaries of higher executives remain comfortable. In 1978 the highest, $398 000, was earned by the president of the Norfolk and Western – a 12 000 km railway and one of the main coal carriers in the US. The figure of $250 000 was exceeded by ten high executives on different networks.

385 The choice of the year 1974 was determined by the economic boom and 1934 by the length of the period.

421 See the report of 16 July 1979 of the Auditor General of the US to Congress: 'The excessive weight of trucks: a burden that we can no longer support'.

421 The reference price of a barrel of 'Arabian light' has increased from $1.80 in 1970 to around $12 between 1973 and 1978, rising on July 1979 to a range of $18.50 to $23.50. The depreciation of the dollar should also be added to this

price escalation. Since the Caracas Conference of OPEC in December 1979 all base reference prices have disappeared.

424 In order to make this survey complete one should mention another mode of transport: electrical energy by high-tension wires. There is no doubt that the location of thermo-electric power stations burning coal or oil has important repercussions on the volume of traffic of railways, waterways and pipelines.

425 By way of example, the TGV of the SNCF, which keeps to the traditional principles of the railway (rail–wheel guidance and steel–steel adhesion) covered 450 000 km in tests at speeds frequently in excess of 300 km/h before the construction of operational equipment which will run at a maximum of 260 km/h.

432 Experiments are being conducted in the US on the RoadRailer III, an improved version of a type of vehicle used 20 years ago. It is a traditional trailer whose twin axles can be removed to give room for a railway axle. These form homogeneous articulated trains similar to the Talgo (see 242). If the experiments are successful, the system could be developed in countries where the difference between the load per axle allowed on the railway and on the road is sufficiently high.

441 The results of the Japanese experiment will be analysed in chapter 6.

511 In France for example, transport capital represents around 25 % of total national capital, and road infrastructures account for 60 % of that figure.

512 Roger Guibert's book *Service public et productivité*, (Société d'édition d'enseignement supérieur, Paris, 1956) has helped considerably to distinguish the essential aspects of the idea of public service in transport.

523 It is to be regretted that, in this respect, some governments submerge transport in a larger structure, such as 'Environment'. This lowers the level of responsibility at a time when public opinion is becoming more and more sensitive to its sound operation.

53 This schema reproduces, in its essentials, one that was presented to the IRCA Congress of Paris in June 1966, within the framework of a report on the use of electronic computers, telecommunications, and the methods of cybernetics in the management of goods traffic.

541 The enormous rise in the price of oil will surely slow down this rate of development.

543 By way of example, the total size of the French railway system decreased from 60 000 km in 1930 to 34 000 km in 1978. It was similar in the Netherlands, where between 1930 and 1937 a quarter of the system was closed and more than half the stations were closed to passenger traffic, making the railway profitable until recent years.

631 'Whatever its origins and its justifications are, railway nationalism now constitutes a handicap which is becoming more and more serious for supply, all the more so as demand is changing quickly and profoundly'. (39th Round Table on the economics of transport of the ECMT, Paris, October 1977).

634 These include the six member countries of the CMEA (German Democratic Republic, Bulgaria, Hungary, Poland, Rumania and Czechoslovakia) as well as Yugoslavia.

663 A plan using wagon-transporter ferries is being studied so as to avoid these two trans-shipments.

671 Joseph R. Daughen and Peter Binzen, *The Wreck of the Penn Central*, Little, Brown and Co, Boston and Toronto, 1971.

672 The transcontinental networks have, however, proved themselves in Canada, as we shall see later on.

732 West Germany, Belgium, Denmark, France, Great Britain, Ireland, Italy, Luxembourg, and the Netherlands, whose railways form the Group of nine in the UIC, which have contacts with the EEC. The admission of Greece in 1981 broke the geographical unity of the EEC. Portugal and Spain are prospective members.

734 See further in 741.

Post. *Message pour ma patrie professionnelle*, edited by the Association des amis de Louis Armand, Paris, 1974.

BIBLIOGRAPHY

There are no recent general works on railways but several engineers have dealt with various aspects of railway technology, and economic problems are often the subject of papers at congresses or seminars. In practice, up to date information must be sought in specialist reviews, reports of congresses and some company reviews. I have listed the most important ones below. Statistical data at railway level used in this book come from the United Nations, government and inter-governmental organisations, and railways. Unfortunately, homogeneous data is not yet produced at world level.

Directories and annual reports

Railway Directory and Year Book, (annually) London.
Jane's World Railways and Rapid Transit Systems, (annually) London.
Facts and Figures (annual reports published by numerous networks and railway authorities).
150th Anniversary of America's Railroads, Yearbook of Railroad Facts (1977) Washington, AAR.
Thomas Cook International Timetable, Railway and Local Shipping Services Guide, (12 times a year) London.
Passenger and goods services timetables from various networks.

Records and technical reports – statements

Records from the 'UIC Code', 10 volumes, 600 leaflets (French, English, German), Paris, UIC.
Reports and technical documents from the Office of Research and Experiment of the UIC (ORE), (more than 900, French, English, German), Utrecht, ORE.
Reports from Symposia on the use of cybernetics on the railway, four collections: Symposium of Paris (1963); Symposium of Montreal (1967); Symposium of Washington (1974); Symposium of Tokyo (1970); (French, English, German), Paris, UIC.
Manual of Standards and Recommended Practices, Washington, AAR.

Research documents and dictionaries

UIC Bulletin of Documentation, (1947–1963, monthly, French and Spanish), Paris, UIC.
Summary of International Railway Documentation IRCA UIC, (1964 onwards, 10 issues a year, French, English, German), Brussels, IRCA, Madrid (Spanish version).
Railroad Research Bulletin, (1974 onwards, six-monthly), Washington, RRIS (Railroad Research Information Service).
UITP Biblio – Index, (1962 onwards, three-monthly, French, German, English, Brussels, UITP (Union Internationale des Transports Publics).
General Glossary of Railway Terms (3rd edition 1975, 11 700 terms, French, German, English, Italian, Spanish, Dutch), Paris, UIC and Karlsruhe, Malsch and Vogel Gmbh.
Ośmiojęzyczy słownik Kolejowy (railway glossary in eight languages), (1978, around 12 000 terms, (French, German, Russian, Hungarian, Polish, Rumanian, Czech, English), Warsaw, Wydawnictwa Komunikacji i ɫTączności.
Railway Thesaurus, (2700 to 3400 terms, French, German, English), Paris, UIC.

Reviews

International reviews

IRCA – Bulletin de l'Association Internationale du Congrès des Chemins de Fer; (1886–1938, 1946–69, monthly), Brussels.
La Traction éléctrique dans les chemins de fer, (1962–February 1964, monthly, Brussels.
Cybernétique et Electronique dans les chemins de fer, (March 1964–69, monthly), Brussels.
UIC – Bulletin de l'Union Internationale des Chemins de Fer, (1925–40, 1946–69, monthly), Paris.
UIC/IRCA – Rail International (French, English); Schienen der Welt; (German), (1970 onwards, monthly), Brussels.
OSZD – Zeitschrift der OSShD, (1957 onwards, bimonthly, German, Russian), Warsaw.
ALAF Revista, (three monthly, Spanish), Buenos Aires.
Boletin de la Comisión Permanente del Asociación del Congreso Panamericano de Ferrocariles, (bimonthly, Spanish), Buenos Aires.
International Railway Journal, (monthly, English), Falmouth.
Internationales Verkehrswesen, (monthly, German), Frankfurt am Main.
Journal pour le transport international, (weekly, French and English), Basle.
Rail Engineering International, (bimonthly, English), London.
Le Rail et le Monde, (monthly, French), Paris (ex. *La Vie du Rail Outre-Mer*).
Railway Engineer International, (bimonthly, English), London.
La Vie du Rail Outre-Mer, 1954–78, monthly, French, Paris.

National reviews

AIT (Asociación de Investigación del Transporte) Revista, (bimonthly, Spanish), Madrid.
AREA (American Railway Engineering Association) Bulletin, (bimonthly, English), Chicago.
Die Bundesbahn, (monthly, German), Darmstadt.
Canadian Transportation and Distribution Management, (monthly, English), Don Mills.
Chemins de fer (AFAC), (monthly, French), Paris.
DDR Verkehr, (monthly, German), Berlin.
Demiryol, (monthly, Turkish), Ankara.
DET – Die Eisenbahntechnik, (monthly, German), Berlin.
DVZ – Deutsche Verkehrs-Zeitung, (tri-weekly, German), Hamburg.
Doprava, (three-monthly, Czech), Prague.
Eisenbahn Technische Rundschau – ETR, (monthly, German), Darmstadt.
Ferrovia e Trasporti, (monthly, Italian), Rome.
Indian Railways, monthly, English), New Delhi.

Ingegneria Ferroviaria, (monthly, Italian), Rome.
Japanese Railway Engineering, (three-monthly, English), Tokyo.
Journal of Advanced Transportation, (three-monthly, English), Durham (US).
Journal of Transport Economics and Policy, (three-monthly, English), London.
Közlekedéstudományi Szemle, (monthly, Hungarian), Budapest.
MR – Modern Railroads/Rail Transit, (monthly, English), Chicago.
Nordisk Järnbane Tidskrift, (Danish, Norwegian, Swedish), Trondheim.
Przeglad Komunikacyjny, (monthly, Polish), Warsaw.
Quarterly Reports of the RTRI (Railway Technical Research Institute, (three-monthly, English), Tokyo.
Railway Age, (bimonthly, English), New York.
Railways of Australia Network, (monthly, English), Sydney.
Revista Ferroviaria, (irregular, Portuguese), Rio de Janeiro.
Revista Transporturilor si Telecommunicatiilor, (monthly, Rumanian), Bucharest.
Revue Générale des Chemins de Fer, (monthly, French), Paris.
Schweizerische Zeitschrift für Verkehrswirtschaft, (three-monthly, German), Vienna.
Vestnik VNIIZT, (8 issues a year, Russian), Moscow.
La Vie de Rail, (weekly, French), Paris.
Zeleznice, (monthly, Serbo-Croat), Belgrade.
Zeleznodoroznyj Transport, (monthly, Russian), Moscow.
Zelezop''ten Transport, (monthly, Bulgarian), Sofia.
ZEV – Glasers Annalen, (monthly, German), Berlin.